JN253870

対訳

ISO 9001:2015

(JIS Q 9001:2015)

ポケット版

品質マネジメントの国際規格

品質マネジメントシステム規格国内委員会　監修

日本規格協会　編

2019年7月1日のJIS法改正により名称が変わりました．本書に収録しているJISについても，まえがきを除き，規格中の「日本工業規格」を「日本産業規格」に読み替えてください．

ISO copyright office
Ch. de Blandonnet 8・CP 401
CH-1214 Vernier, Geneva, Switzerland
Tel. +41 22 749 01 11
Fax +41 22 749 09 47
E-mail copyright@iso.org
Web www.iso.org

はじめに

品質保証の国際規格として1987年に制定されたISO 9001は，1994年，2000年，2008年と改訂を重ね2015年には第5版となるISO 9001:2015（Quality management systems — Requirements）が発行された。

このISO 9001:2015では，“ISO/IEC専門業務用指針—第1部：統合版ISO補足指針”の附属書SLへの対応を大きな柱とした改訂が行われた。これにより箇条構成や要求事項の表現が変更されるとともに，品質マネジメントシステムの事業プロセスへの統合や，リスク及び機会への取組み，サービス分野へのいっそうの配慮が取り入れられるなど，大規模な改訂となっている。

国内においては，品質マネジメントシステム規格国内委員会を中心にISO 9001対応WGが設置され，ISO 9001:2015を翻訳したJIS Q 9001（品質マネジメントシステム—要求事項）が同じく2015年に改正された。

また，ISO 9001:2015で引用されるISO 9000（Quality management systems — Fundamentals and vocabulary）も 2015 年に改訂され，その完全一致の翻訳規格である JIS Q 9000（品質マネジメントシステム―基本及び用語）も 2015 年に改正されている。

本書は，この ISO と JIS を英和対訳版として発刊するものである（ISO/JIS Q 9000 は一部抜粋）。

英文は ISO 9001:2015 及び ISO 9000:2015 を ISO（国際標準化機構）の許可を得て，収録している。また，和文は，それぞれの ISO につき，技術的内容及び対応国際規格の構成を変更することなく作成され，日本工業標準調査会で審議を経て改正された JIS Q 9001:2015 及び JIS Q 9000:2015 である。なお，JIS 規格文中において点線下線を施した箇所は，原国際規格にはない事項であること，また，JIS の解説は省略していることに留意いただきたい。

このたびの ISO 9001:2015，ISO 9000:2015 の規格開発，及び JIS Q 9001，JIS Q 9000 の 2015 年改正にあたり，品質マネジメントシステム規格国内委員会委員の方々には，国内外の会議での検討など，多大なご尽力をいただいた。ここに厚く感謝の意を表する。

本書が，品質マネジメントシステムの認証の更新や維持，あるいはこれから新たに審査登録を受ける企業のための原典として，前版同様お役に立てば幸いである。

また，これらの規格をより深く理解したい方に対しては，書籍『ISO 9001:2015 要求事項の解説』，『ISO 9001:2015 新旧規格の対照と解説』（いずれも，日本規格協会，2015）をはじめとする関係書籍を併読されることをお勧めする。

2016 年 2 月

日本規格協会

品質マネジメントシステム規格国内委員会

（2015 年 11 月現在）

［委員長］	中條　武志	中央大学
［副委員長］	棟近　雅彦	早稲田大学
	山田　　秀	筑波大学
［委員］	相澤　　滋	一般社団法人情報通信ネットワーク産業協会（株式会社日立製作所）
	秋山　文生	一般社団法人日本建設業連合会（清水建設株式会社）
	浅井　英規	ISO/TC 210 国内対策委員会（株式会社日立ハイテクサイエンス）
	足立　憲昭	イオンエンターテイメント株式会社
	阿部　　隆	一般社団法人日本鉄鋼連盟
	安藤　之裕	合資会社安藤技術士事務所
	飯尾　隆弘	日本マネジメントシステム認証機関協議会（ロイド レジスター クオリティ アシュアランス リミテッド）
	飯塚　悦功	東京大学
	岩岡　　覚	電気事業連合会（～ H27.6.30）
	今木　　圭	電気事業連合会（H27.7.1 ～）
	上中　浩幸	一般財団法人日本規格協会 マネジメントシステム審査員評価登録センター
	尾島　善一	ISO/TC 69 本委員会（東京理科大学）
	小野　　哲	一般社団法人産業機械工業会（株式会社荏原製作所）
	菅野　義就	一般社団法人日本航空宇宙工業会（～ H27.7.31）
	前畑　貴芳	一般社団法人日本航空宇宙工業会（H27.8.1 ～）

	久米　　均	東京大学
	黒田　晋一	一般社団法人日本計量振興協会 (株式会社島津製作所)
	小菅　雄治	東京商工会議所 (日本システムウエア株式会社)
	芝原　　純	公益社団法人消費者関連専門家会議
	須田　晋介	株式会社テクノファ
	住本　　守	元 ソニー株式会社
	武樋　憲明	審査員研修機関連絡協議会 (一般社団法人日本能率協会)
	塚原　敏郎	一般社団法人電子情報技術産業協会 (三菱電機株式会社) (～ H27.3.31)
	宮下　正則	一般社団法人電子情報技術産業協会 (富士通株式会社) (H27.4.1 ～)
	水流　聡子	東京大学大学院
	中川　　梓	公益財団法人日本適合性認定協会
	中村　公美	一般社団法人人材育成と教育サービス協議会
	福田　泰和	経済産業省
	福丸　典芳	有限会社福丸マネジメントテクノ
	松本　芳彦	一般社団法人日本化学工業協会
	三浦　重孝	サクラ精機株式会社
	吉田　孝一	一般社団法人日本電機工業会
	吉田　敬史	ISO/TC 207/SC 1 小委員会 (合同会社グリーンフューチャーズ)
[関係者]	藤代　尚武	経済産業省
	宮尾　　健	経済産業省
	岡崎　　将	経済産業省
[事務局]	平岡　靖敏	一般財団法人日本規格協会
	古野　　毅	一般財団法人日本規格協会
	佐藤　恭子	一般財団法人日本規格協会
	諸橋　護易	一般財団法人日本規格協会
	鹿山　優子	一般財団法人日本規格協会
	高井　玉歩	一般財団法人日本規格協会

(敬称略)

ISO 9001 対応 WG

（2015 年 11 月現在）

［主査］	中條　武志	中央大学
［委員］	朝倉　崇顕	株式会社 IHI
	足立　憲昭	イオンエンターテイメント株式会社
	飯塚　悦功	東京大学
	石川　厚史	新日鐵住金株式会社
	国府　保周	活き活き経営システムズ
	須田　晋介	株式会社テクノファ
	住本　　守	元 ソニー株式会社
	中川　　梓	公益財団法人日本適合性認定協会
	平林　良人	株式会社テクノファ
	棟近　雅彦	早稲田大学
	山田　　秀	筑波大学
	吉田　敬史	合同会社グリーンフューチャーズ
	米岡　優子	ロイド レジスター クオリティ アシュアランス リミテッド
	渡邉　邦道	一般社団法人原子力安全推進協会
［関係者］	宮尾　　健	経済産業省
	岡崎　　将	経済産業省
［事務局］	佐藤　恭子	一般財団法人日本規格協会
	諸橋　護易	一般財団法人日本規格協会
	鹿山　優子	一般財団法人日本規格協会
	高井　玉歩	一般財団法人日本規格協会

（敬称略）

ISO 9000 対応 WG

(2015 年 11 月現在)

[主査]	中條　武志	中央大学
[委員]	安藤　之裕	合資会社安藤技術士事務所
	飯塚　悦功	東京大学
	須田　晋介	株式会社テクノファ
	住本　　守	元 ソニー株式会社
	水流　聡子	東京大学大学院
	平林　良人	株式会社テクノファ
	棟近　雅彦	早稲田大学
	山田　　秀	筑波大学
[関係者]	宮尾　　健	経済産業省
	岡崎　　将	経済産業省
[事務局]	佐藤　恭子	一般財団法人日本規格協会
	諸橋　護易	一般財団法人日本規格協会
	鹿山　優子	一般財団法人日本規格協会
	高井　玉歩	一般財団法人日本規格協会

(敬称略)

Contents

ISO 9001:2015
Quality management systems — Requirements

目　次

JIS Q 9001:2015
品質マネジメントシステム
―要求事項

ISO 9000:2015
Quality management systems — Fundamentals and vocabulary（抜粋）

JIS Q 9000:2015
品質マネジメントシステム
—基本及び用語（抜粋）

ISO 9001
Fifth edition 2015-9-15

JIS Q 9001
2015-11-20

Quality management systems —Requirements

品質マネジメントシステム —要求事項

Foreword

ISO (the International Organization for Standardization) is a worldwide federation of national standards bodies (ISO member bodies). The work of preparing International Standards is normally carried out through ISO technical committees. Each member body interested in a subject for which a technical committee has been established has the right to be represented on that committee. International organizations, governmental and non-governmental, in liaison with ISO, also take part in the work. ISO collaborates closely with the International Electrotechnical Commission (IEC) on all matters of electrotechnical standardization.

The procedures used to develop this document and those intended for its further maintenance are described in the ISO/IEC Directives, Part 1. In particular the different approval criteria needed for the different types of ISO documents should be noted. This document was drafted in

まえがき

（ISO の Foreword と JIS のまえがきは，それぞれの原文において内容が異なっているため，対訳となっていないことにご注意ください.）

この規格は，工業標準化法第 14 条によって準用する第 12 条第 1 項の規定に基づき，一般財団法人日本規格協会（JSA）から，工業標準原案を具して日本工業規格を改正すべきとの申出があり，日本工業標準調査会の審議を経て，経済産業大臣が改正した日本工業規格である．これによって，**JIS Q 9001**:2008 は改正され，この規格に置き換えられた．

この規格は，著作権法で保護対象となっている著作物である．

この規格の一部が，特許権，出願公開後の特許出願又は実用新案権に抵触する可能性があることに注意を喚起する．経済産業大臣及び日本工業標準調査会は，このような特許権，出願公開後の特許出願及び実用新案権に関わる確認について，責任はもたない．

accordance with the editorial rules of the ISO/IEC Directives, Part 2 (see www.iso.org/directives).

Attention is drawn to the possibility that some of the elements of this document may be the subject of patent rights. ISO shall not be held responsible for identifying any or all such patent rights. Details of any patent rights identified during the development of the document will be in the Introduction and/or on the ISO list of patent declarations received (see www.iso.org/patents).

Any trade name used in this document is information given for the convenience of users and does not constitute an endorsement.

For an explanation on the meaning of ISO specific terms and expressions related to conformity assessment, as well as information about ISO's adherence to the World Trade Organization (WTO) principles in the Technical Barriers to Trade (TBT) see the following URL: www.iso.org/iso/foreword.html.

The committee responsible for this document is Technical Committee ISO/TC 176, *Quality management and quality assurance,* Subcommittee SC 2, *Quality systems.*

This fifth edition cancels and replaces the fourth edition (ISO 9001:2008), which has been technically revised, through the adoption of a revised clause sequence and the adaptation of the revised quality management principles and of new concepts. It also cancels and replaces the Technical Corrigendum ISO 9001:2008/Cor.1:2009.

Introduction

0.1 General

The adoption of a quality management system is a strategic decision for an organization that can help to improve its overall performance and provide a sound basis for sustainable development initiatives.

The potential benefits to an organization of implementing a quality management system based on this International Standard are:

a) the ability to consistently provide products and services that meet customer and applicable statutory and regulatory requirements;

b) facilitating opportunities to enhance customer satisfaction;

c) addressing risks and opportunities associated with its context and objectives;

序文

この規格は，2015年に第5版として発行された**ISO 9001**を基に，技術的内容及び構成を変更することなく作成した日本工業規格である．

なお，この規格で点線の下線を施してある参考事項は，対応国際規格にはない事項である．

0.1　一般

品質マネジメントシステムの採用は，パフォーマンス全体を改善し，持続可能な発展への取組みのための安定した基盤を提供するのに役立ち得る，組織の戦略上の決定である．

組織は，この規格に基づいて品質マネジメントシステムを実施することで，次のような便益を得る可能性がある．

a)　顧客要求事項及び適用される法令・規制要求事項を満たした製品及びサービスを一貫して提供できる．

b)　顧客満足を向上させる機会を増やす．

c)　組織の状況及び目標に関連したリスク及び機会に取り組む．

d) the ability to demonstrate conformity to specified quality management system requirements.

This International Standard can be used by internal and external parties.

It is not the intent of this International Standard to imply the need for:

— uniformity in the structure of different quality management systems;
— alignment of documentation to the clause structure of this International Standard;
— the use of the specific terminology of this International Standard within the organization.

The quality management system requirements specified in this International Standard are complementary to requirements for products and services.

This International Standard employs the process approach, which incorporates the Plan-Do-Check-Act (PDCA) cycle and risk-based thinking.

d)　規定された品質マネジメントシステム要求事項への適合を実証できる.

内部及び外部の関係者がこの規格を使用することができる.

この規格は, 次の事項の必要性を示すことを意図したものではない.

— 様々な品質マネジメントシステムの構造を画一化する.

— 文書類をこの規格の箇条の構造と一致させる.

— この規格の特定の用語を組織内で使用する.

この規格で規定する品質マネジメントシステム要求事項は, 製品及びサービスに関する要求事項を補完するものである.

この規格は, Plan-Do-Check-Act (PDCA) サイクル及びリスクに基づく考え方を組み込んだ, プロセスアプローチを用いている.

The process approach enables an organization to plan its processes and their interactions.

The PDCA cycle enables an organization to ensure that its processes are adequately resourced and managed, and that opportunities for improvement are determined and acted on.

Risk-based thinking enables an organization to determine the factors that could cause its processes and its quality management system to deviate from the planned results, to put in place preventive controls to minimize negative effects and to make maximum use of opportunities as they arise (see Clause A.4).

Consistently meeting requirements and addressing future needs and expectations poses a challenge for organizations in an increasingly dynamic and complex environment. To achieve this objective, the organization might find it necessary to adopt various forms of improvement in addition to correction and continual improvement, such as breakthrough change, innovation and re-

組織は，プロセスアプローチによって，組織のプロセス及びそれらの相互作用を計画することができる．

組織は，PDCA サイクルによって，組織のプロセスに適切な資源を与え，マネジメントすることを確実にし，かつ，改善の機会を明確にし，取り組むことを確実にすることができる．

組織は，リスクに基づく考え方によって，自らのプロセス及び品質マネジメントシステムが，計画した結果からかい（乖）離することを引き起こす可能性のある要因を明確にすることができ，また，好ましくない影響を最小限に抑えるための予防的管理を実施することができ，更に機会が生じたときにそれを最大限に利用することができる（**A.4** 参照）．

ますます動的で複雑になる環境において，一貫して要求事項を満たし，将来のニーズ及び期待に取り組むことは，組織にとって容易ではない．組織は，この目標を達成するために，修正及び継続的改善に加えて，飛躍的な変化，革新，組織再編など様々な改善の形を採用する必要があることを見出すであろう．

organization.

In this International Standard, the following verbal forms are used:

— “shall” indicates a requirement;

— “should” indicates a recommendation;

— “may” indicates a permission;

— “can” indicates a possibility or a capability.

Information marked as “NOTE” is for guidance in understanding or clarifying the associated requirement.

0.2 Quality management principles

This International Standard is based on the quality management principles described in ISO 9000. The descriptions include a statement of each principle, a rationale of why the principle is important for the organization, some examples of benefits associated with the principle and examples of typical actions to improve the organization's performance when applying the principle.

この規格では，次のような表現形式を用いている．

— “～しなければならない”（shall）は，要求事項を示し，
— “～することが望ましい”（should）は，推奨を示し，
— “～してもよい”（may）は，許容を示し，
— “～することができる”，“～できる”，“～し得る”など（can）は，可能性又は実現能力を示す．

“注記”に記載されている情報は，関連する要求事項の内容を理解するための，又は明解にするための手引である．

0.2　品質マネジメントの原則

この規格は，**JIS Q 9000** に規定されている品質マネジメントの原則に基づいている．この規定には，それぞれの原則の説明，組織にとって原則が重要であることの根拠，原則に関連する便益の例，及び原則を適用するときに組織のパフォーマンスを改善するための典型的な取組みの例が含まれている．

The quality management principles are:

— customer focus;
— leadership;
— engagement of people;
— process approach;
— improvement;
— evidence-based decision making;
— relationship management.

0.3 Process approach

0.3.1 General

This International Standard promotes the adoption of a process approach when developing, implementing and improving the effectiveness of a quality management system, to enhance customer satisfaction by meeting customer requirements. Specific requirements considered essential to the adoption of a process approach are included in 4.4.

Understanding and managing interrelated processes as a system contributes to the organization's effectiveness and efficiency in achieving its intended results. This approach enables the organization to control the interrelationships and in-

品質マネジメントの原則とは，次の事項をいう．

— 顧客重視
— リーダーシップ
— 人々の積極的参加
— プロセスアプローチ
— 改善
— 客観的事実に基づく意思決定
— 関係性管理

0.3　プロセスアプローチ

0.3.1　一般

この規格は，顧客要求事項を満たすことによって顧客満足を向上させるために，品質マネジメントシステムを構築し，実施し，その品質マネジメントシステムの有効性を改善する際に，プロセスアプローチを採用することを促進する．プロセスアプローチの採用に不可欠と考えられる特定の要求事項を **4.4** に規定している．

システムとして相互に関連するプロセスを理解し，マネジメントすることは，組織が効果的かつ効率的に意図した結果を達成する上で役立つ．組織は，このアプローチによって，システムのプロセス間の相互関係及び相互依存性を管理することができ，そ

terdependencies among the processes of the system, so that the overall performance of the organization can be enhanced.

The process approach involves the systematic definition and management of processes, and their interactions, so as to achieve the intended results in accordance with the quality policy and strategic direction of the organization. Management of the processes and the system as a whole can be achieved using the PDCA cycle (see 0.3.2) with an overall focus on risk-based thinking (see 0.3.3) aimed at taking advantage of opportunities and preventing undesirable results.

The application of the process approach in a quality management system enables:

a) understanding and consistency in meeting requirements;
b) the consideration of processes in terms of added value;
c) the achievement of effective process performance;
d) improvement of processes based on evalua-

れによって，組織の全体的なパフォーマンスを向上させることができる．

プロセスアプローチは，組織の品質方針及び戦略的な方向性に従って意図した結果を達成するために，プロセス及びその相互作用を体系的に定義し，マネジメントすることに関わる．PDCA サイクル（**0.3.2** 参照）を，機会の利用及び望ましくない結果の防止を目指すリスクに基づく考え方（**0.3.3** 参照）に全体的な焦点を当てて用いることで，プロセス及びシステム全体をマネジメントすることができる．

品質マネジメントシステムでプロセスアプローチを適用すると，次の事項が可能になる．

a)　要求事項の理解及びその一貫した充足

b)　付加価値の点からの，プロセスの検討

c)　効果的なプロセスパフォーマンスの達成

d)　データ及び情報の評価に基づく，プロセスの改善

tion of data and information.

Figure 1 gives a schematic representation of any process and shows the interaction of its elements. The monitoring and measuring check points, which are necessary for control, are specific to each process and will vary depending on the related risks.

0.3.2 Plan-Do-Check-Act cycle

The PDCA cycle can be applied to all processes and to the quality management system as a whole. Figure 2 illustrates how Clauses 4 to 10 can be grouped in relation to the PDCA cycle.

図 1 は，プロセスを図示し，その要素の相互作用を示したものである．管理のために必要な，監視及び測定のチェックポイントは，各プロセスに固有なものであり，関係するリスクによって異なる．

0.3.2　PDCA サイクル

PDCA サイクルは，あらゆるプロセス及び品質マネジメントシステム全体に適用できる．**図 2** は，箇条 **4** ～箇条 **10** を PDCA サイクルとの関係でどのようにまとめることができるかを示したものである．

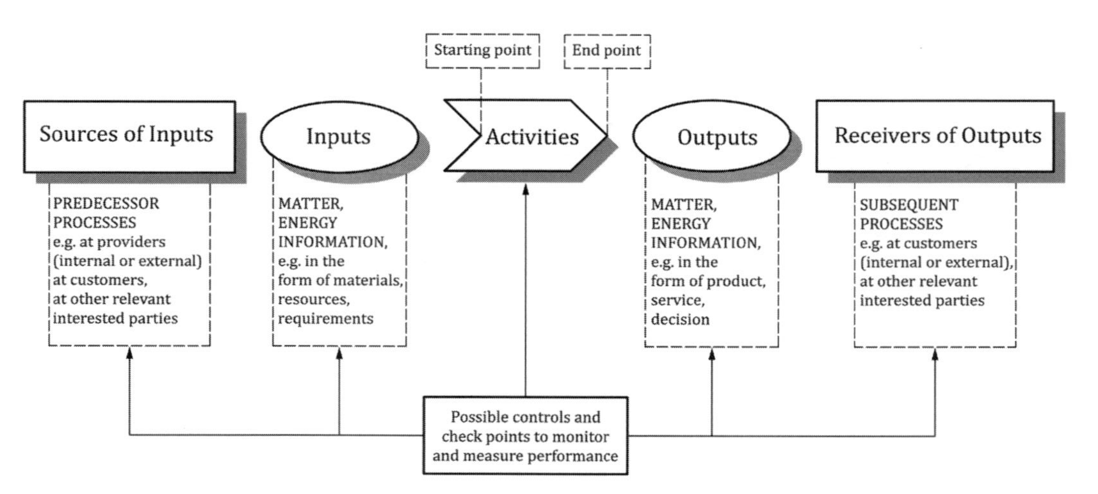

Figure 1 — Schematic representation of the elements of a single process

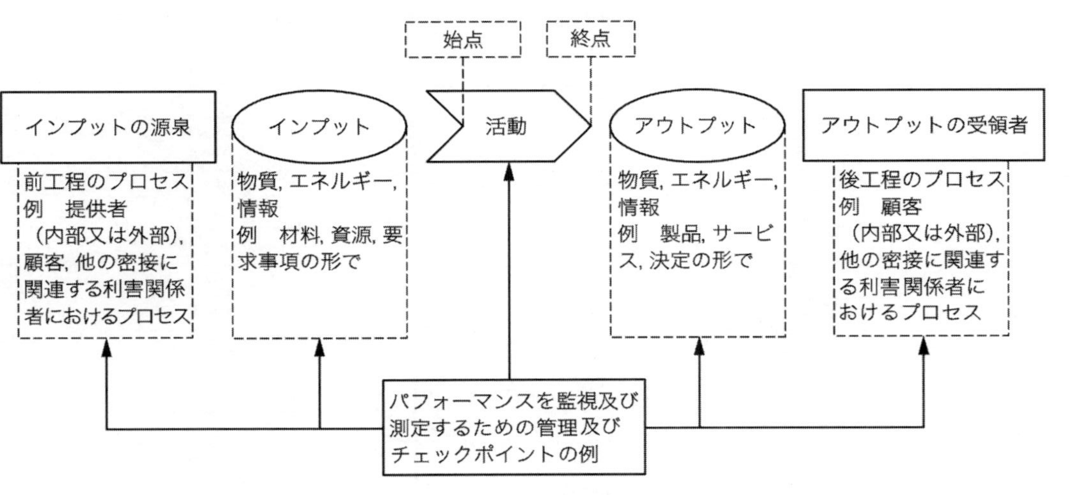

図 1—単一プロセスの要素の図示

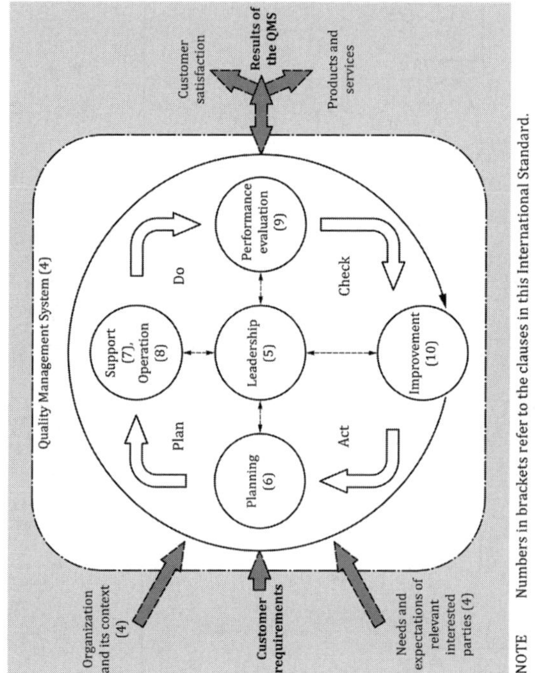

NOTE Numbers in brackets refer to the clauses in this International Standard.

Figure 2 — Representation of the structure of this International Standard in the PDCA cycle

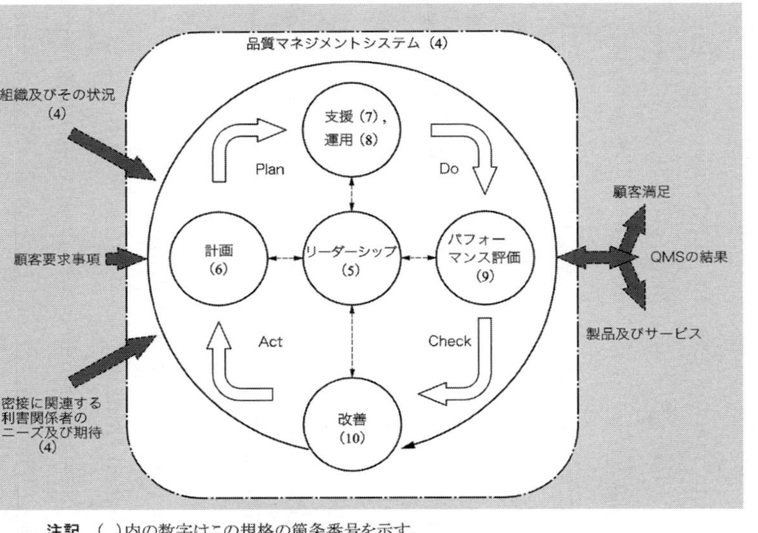

注記　（ ）内の数字はこの規格の箇条番号を示す。

図 2 — PDCA サイクルを使った，この規格の構造の説明

The PDCA cycle can be briefly described as follows:

— **Plan**: establish the objectives of the system and its processes, and the resources needed to deliver results in accordance with customers' requirements and the organization's policies, and identify and address risks and opportunities;

— **Do**: implement what was planned;

— **Check**: monitor and (where applicable) measure processes and the resulting products and services against policies, objectives, requirements and planned activities, and report the results;

— **Act**: take actions to improve performance, as necessary.

0.3.3 Risk-based thinking

Risk-based thinking (see Clause A.4) is essential for achieving an effective quality management system. The concept of risk-based thinking has been implicit in previous editions of this International Standard including, for example, carrying out preventive action to eliminate potential nonconformities, analysing any nonconformities that

PDCA サイクルは，次のように簡潔に説明できる．

— Plan：システム及びそのプロセスの目標を設定し，顧客要求事項及び組織の方針に沿った結果を出すために必要な資源を用意し，リスク及び機会を特定し，かつ，それらに取り組む．

— Do：計画されたことを実行する．

— Check：方針，目標，要求事項及び計画した活動に照らして，プロセス並びにその結果としての製品及びサービスを監視し，（該当する場合には，必ず）測定し，その結果を報告する．

— Act：必要に応じて，パフォーマンスを改善するための処置をとる．

0.3.3　リスクに基づく考え方

リスクに基づく考え方（**A.4** 参照）は，有効な品質マネジメントシステムを達成するために必須である．リスクに基づく考え方の概念は，例えば，起こり得る不適合を除去するための予防処置を実施する，発生したあらゆる不適合を分析する，及び不適合の影響に対して適切な，再発防止のための取組みを行うということを含めて，この規格の旧版に含ま

do occur, and taking action to prevent recurrence that is appropriate for the effects of the nonconformity.

To conform to the requirements of this International Standard, an organization needs to plan and implement actions to address risks and opportunities. Addressing both risks and opportunities establishes a basis for increasing the effectiveness of the quality management system, achieving improved results and preventing negative effects.

Opportunities can arise as a result of a situation favourable to achieving an intended result, for example, a set of circumstances that allow the organization to attract customers, develop new products and services, reduce waste or improve productivity. Actions to address opportunities can also include consideration of associated risks. Risk is the effect of uncertainty and any such uncertainty can have positive or negative effects. A positive deviation arising from a risk can provide an opportunity, but not all positive effects of risk

れていた.

組織は，この規格の要求事項に適合するために，リスク及び機会への取組みを計画し，実施する必要がある. リスク及び機会の双方への取組みによって，品質マネジメントシステムの有効性の向上，改善された結果の達成，及び好ましくない影響の防止のための基礎が確立する.

機会は，意図した結果を達成するための好ましい状況，例えば，組織が顧客を引き付け，新たな製品及びサービスを開発し，無駄を削減し，又は生産性を向上させることを可能にするような状況の集まりの結果として生じることがある. 機会への取組みには，関連するリスクを考慮することも含まれ得る. リスクとは，不確かさの影響であり，そうした不確かさは，好ましい影響又は好ましくない影響をもち得る. リスクから生じる，好ましい方向へのかい(乖)離は，機会を提供し得るが，リスクの好ましい影響の全てが機会をもたらすとは限らない.

result in opportunities.

0.4 Relationship with other management system standards

This International Standard applies the framework developed by ISO to improve alignment among its International Standards for management systems (see Clause A.1).

This International Standard enables an organization to use the process approach, coupled with the PDCA cycle and risk-based thinking, to align or integrate its quality management system with the requirements of other management system standards.

This International Standard relates to ISO 9000 and ISO 9004 as follows:

— ISO 9000 *Quality management systems — Fundamentals and vocabulary* provides essential background for the proper understanding and implementation of this International Standard;

— ISO 9004 *Managing for the sustained success*

0.4　他のマネジメントシステム規格との関係

この規格は，マネジメントシステムに関する規格間の一致性を向上させるために国際標準化機構（**ISO**）が作成した枠組みを適用する（**A.1** 参照）．

この規格は，組織が，品質マネジメントシステムを他のマネジメントシステム規格の要求事項に合わせたり，又は統合したりするために，PDCA サイクル及びリスクに基づく考え方と併せてプロセスアプローチを用いることができるようにしている．

この規格は，次に示す **JIS Q 9000** 及び **JIS Q 9004** に関係している．

— **JIS Q 9000**（品質マネジメントシステム—基本及び用語）は，この規格を適切に理解し，実施するために不可欠な予備知識を与えている．

— **JIS Q 9004**（組織の持続的成功のための運営

of an organization — A quality management approach provides guidance for organizations that choose to progress beyond the requirements of this International Standard.

Annex B provides details of other International Standards on quality management and quality management systems that have been developed by ISO/TC 176.

This International Standard does not include requirements specific to other management systems, such as those for environmental management, occupational health and safety management, or financial management.

Sector-specific quality management system standards based on the requirements of this International Standard have been developed for a number of sectors. Some of these standards specify additional quality management system requirements, while others are limited to providing guidance to the application of this International Standard within the particular sector.

管理—品質マネジメントアプローチ）は，この規格の要求事項を超えて進んでいくことを選択する組織のための手引を提供している．

附属書 B は，**ISO/TC 176** が作成した他の品質マネジメント及び品質マネジメントシステム規格類について詳述している．

この規格には，環境マネジメント，労働安全衛生マネジメント又は財務マネジメントのような他のマネジメントシステムに固有な要求事項は含んでいない．

幾つかの分野において，この規格の要求事項に基づく，分野固有の品質マネジメントシステム規格が作成されている．これらの規格の中には，品質マネジメントシステムの追加的な要求事項を規定しているものもあれば，特定の分野内でのこの規格の適用に関する手引の提供に限定しているものもある．

A matrix showing the correlation between the clauses of this edition of this International Standard and the previous edition (ISO 9001:2008) can be found on the ISO/TC 176/SC 2 open access web site at: www.iso.org/tc176/sc02/public.

1 Scope

This International Standard specifies requirements for a quality management system when an organization:

a) needs to demonstrate its ability to consistently provide products and services that meet customer and applicable statutory and regulatory requirements, and

b) aims to enhance customer satisfaction through the effective application of the system, including processes for improvement of the system and the assurance of conformity to customer and applicable statutory and regulatory requirements.

All the requirements of this International Standard are generic and are intended to be applicable to any organization, regardless of its type or

この規格が基礎とした**ISO 9001**:2015と旧版（**ISO 9001**:2008）との間の箇条の相関に関するマトリクスは，**ISO/TC 176/SC 2**のウェブサイト（www.iso.org/tc176/sc02/public）で公表されている．

1　適用範囲

この規格は，次の場合の品質マネジメントシステムに関する要求事項について規定する．

a)　組織が，顧客要求事項及び適用される法令・規制要求事項を満たした製品及びサービスを一貫して提供する能力をもつことを実証する必要がある場合．

b)　組織が，品質マネジメントシステムの改善のプロセスを含むシステムの効果的な適用，並びに顧客要求事項及び適用される法令・規制要求事項への適合の保証を通して，顧客満足の向上を目指す場合．

この規格の要求事項は，汎用性があり，業種・形態，規模，又は提供する製品及びサービスを問わず，あらゆる組織に適用できることを意図している．

size, or the products and services it provides.

NOTE 1 In this International Standard, the terms "product" or "service" only apply to products and services intended for, or required by, a customer.

NOTE 2 Statutory and regulatory requirements can be expressed as legal requirements.

2 Normative references

The following documents, in whole or in part, are normatively referenced in this document and are indispensable for its application. For dated references, only the edition cited applies. For undated references, the latest edition of the referenced document (including any amendments) applies.

注記 1　この規格の“製品”又は“サービス”という用語は，顧客向けに意図した製品及びサービス，又は顧客に要求された製品及びサービスに限定して用いる．

注記 2　法令・規制要求事項は，法的要求事項と表現することもある．

注記 3　この規格の対応国際規格及びその対応の程度を表す記号を，次に示す．

ISO 9001:2015，Quality management systems—Requirements（IDT）

なお，対応の程度を表す記号“IDT”は，**ISO/IEC Guide 21-1** に基づき，“一致している”ことを示す．

2　引用規格

次に掲げる規格は，この規格に引用されることによって，この規格の規定の一部を構成する．この引用規格は，記載の年の版を適用し，その後の改正版（追補を含む．）は適用しない．

ISO 9000:2015, *Quality management systems — Fundamentals and vocabulary*

3 Terms and definitions

For the purposes of this document, the terms and definitions given in ISO 9000:2015 apply.

4 Context of the organization

4.1 Understanding the organization and its context

The organization shall determine external and internal issues that are relevant to its purpose and its strategic direction and that affect its ability to achieve the intended result(s) of its quality management system.

The organization shall monitor and review information about these external and internal issues.

NOTE 1 Issues can include positive and negative factors or conditions for consideration.

JIS Q 9000:2015　品質マネジメントシステム—基本及び用語

注記　対応国際規格：**ISO 9000**:2015, Quality management systems—Fundamentals and vocabulary (IDT)

3　用語及び定義

この規格で用いる主な用語及び定義は，**JIS Q 9000**:2015 による．

4　組織の状況

4.1　組織及びその状況の理解

組織は，組織の目的及び戦略的な方向性に関連し，かつ，その品質マネジメントシステムの意図した結果を達成する組織の能力に影響を与える，外部及び内部の課題を明確にしなければならない．

組織は，これらの外部及び内部の課題に関する情報を監視し，レビューしなければならない．

注記 1　課題には，検討の対象となる，好ましい要因又は状態，及び好ましくない要

NOTE 2 Understanding the external context can be facilitated by considering issues arising from legal, technological, competitive, market, cultural, social and economic environments, whether international, national, regional or local.

NOTE 3 Understanding the internal context can be facilitated by considering issues related to values, culture, knowledge and performance of the organization.

4.2 Understanding the needs and expectations of interested parties

Due to their effect or potential effect on the organization's ability to consistently provide products and services that meet customer and applicable statutory and regulatory requirements, the organization shall determine:

a) the interested parties that are relevant to the quality management system;
b) the requirements of these interested parties that are relevant to the quality management system.

因又は状態が含まれ得る.

注記 2　外部の状況の理解は, 国際, 国内, 地方又は地域を問わず, 法令, 技術, 競争, 市場, 文化, 社会及び経済の環境から生じる課題を検討することによって容易になり得る.

注記 3　内部の状況の理解は, 組織の価値観, 文化, 知識及びパフォーマンスに関する課題を検討することによって容易になり得る.

4.2　利害関係者のニーズ及び期待の理解

次の事項は, 顧客要求事項及び適用される法令・規制要求事項を満たした製品及びサービスを一貫して提供する組織の能力に影響又は潜在的影響を与えるため, 組織は, これらを明確にしなければならない.

a)　品質マネジメントシステムに密接に関連する利害関係者

b)　品質マネジメントシステムに密接に関連するそれらの利害関係者の要求事項

The organization shall monitor and review information about these interested parties and their relevant requirements.

4.3 Determining the scope of the quality management system

The organization shall determine the boundaries and applicability of the quality management system to establish its scope.

When determining this scope, the organization shall consider:

a) the external and internal issues referred to in 4.1;
b) the requirements of relevant interested parties referred to in 4.2;
c) the products and services of the organization.

The organization shall apply all the requirements of this International Standard if they are applicable within the determined scope of its quality management system.

The scope of the organization's quality manage-

組織は，これらの利害関係者及びその関連する要求事項に関する情報を監視し，レビューしなければならない．

4.3　品質マネジメントシステムの適用範囲の決定

組織は，品質マネジメントシステムの適用範囲を定めるために，その境界及び適用可能性を決定しなければならない．

この適用範囲を決定するとき，組織は，次の事項を考慮しなければならない．

a)　**4.1** に規定する外部及び内部の課題

b)　**4.2** に規定する，密接に関連する利害関係者の要求事項

c)　組織の製品及びサービス

決定した品質マネジメントシステムの適用範囲内でこの規格の要求事項が適用可能ならば，組織は，これらを全て適用しなければならない．

組織の品質マネジメントシステムの適用範囲は，

ment system shall be available and be maintained as documented information. The scope shall state the types of products and services covered, and provide justification for any requirement of this International Standard that the organization determines is not applicable to the scope of its quality management system.

Conformity to this International Standard may only be claimed if the requirements determined as not being applicable do not affect the organization's ability or responsibility to ensure the conformity of its products and services and the enhancement of customer satisfaction.

4.4 Quality management system and its processes

4.4.1 The organization shall establish, implement, maintain and continually improve a quality management system, including the processes needed and their interactions, in accordance with the requirements of this International Standard.

The organization shall determine the processes

文書化した情報として利用可能な状態にし，維持しなければならない．適用範囲では，対象となる製品及びサービスの種類を明確に記載し，組織が自らの品質マネジメントシステムの適用範囲への適用が不可能であることを決定したこの規格の要求事項全てについて，その正当性を示さなければならない．

適用不可能なことを決定した要求事項が，組織の製品及びサービスの適合並びに顧客満足の向上を確実にする組織の能力又は責任に影響を及ぼさない場合に限り，この規格への適合を表明してよい．

4.4　品質マネジメントシステム及びそのプロセス

4.4.1　組織は，この規格の要求事項に従って，必要なプロセス及びそれらの相互作用を含む，品質マネジメントシステムを確立し，実施し，維持し，かつ，継続的に改善しなければならない．

組織は，品質マネジメントシステムに必要なプロ

needed for the quality management system and their application throughout the organization, and shall:

a) determine the inputs required and the outputs expected from these processes;

b) determine the sequence and interaction of these processes;

c) determine and apply the criteria and methods (including monitoring, measurements and related performance indicators) needed to ensure the effective operation and control of these processes;

d) determine the resources needed for these processes and ensure their availability;

e) assign the responsibilities and authorities for these processes;

f) address the risks and opportunities as determined in accordance with the requirements of 6.1;

g) evaluate these processes and implement any changes needed to ensure that these processes achieve their intended results;

h) improve the processes and the quality man-

セス及びそれらの組織全体にわたる適用を決定しなければならない．また，次の事項を実施しなければならない．

a)　これらのプロセスに必要なインプット，及びこれらのプロセスから期待されるアウトプットを明確にする．

b)　これらのプロセスの順序及び相互作用を明確にする．

c)　これらのプロセスの効果的な運用及び管理を確実にするために必要な判断基準及び方法（監視，測定及び関連するパフォーマンス指標を含む．）を決定し，適用する．

d)　これらのプロセスに必要な資源を明確にし，及びそれが利用できることを確実にする．

e)　これらのプロセスに関する責任及び権限を割り当てる．

f)　**6.1** の要求事項に従って決定したとおりにリスク及び機会に取り組む．

g)　これらのプロセスを評価し，これらのプロセスの意図した結果の達成を確実にするために必要な変更を実施する．

h)　これらのプロセス及び品質マネジメントシステ

agement system.

4.4.2 To the extent necessary, the organization shall:

a) maintain documented information to support the operation of its processes;

b) retain documented information to have confidence that the processes are being carried out as planned.

5 Leadership

5.1 Leadership and commitment

5.1.1 General

Top management shall demonstrate leadership and commitment with respect to the quality management system by:

a) taking accountability for the effectiveness of the quality management system;

b) ensuring that the quality policy and quality objectives are established for the quality management system and are compatible with the context and strategic direction of the organization;

c) ensuring the integration of the quality man-

ムを改善する.

4.4.2　組織は，必要な程度まで，次の事項を行わなければならない.

a)　プロセスの運用を支援するための文書化した情報を維持する.

b)　プロセスが計画どおりに実施されたと確信するための文書化した情報を保持する.

5　リーダーシップ

5.1　リーダーシップ及びコミットメント

5.1.1　一般

トップマネジメントは，次に示す事項によって，品質マネジメントシステムに関するリーダーシップ及びコミットメントを実証しなければならない.

a)　品質マネジメントシステムの有効性に説明責任（accountability）を負う.

b)　品質マネジメントシステムに関する品質方針及び品質目標を確立し，それらが組織の状況及び戦略的な方向性と両立することを確実にする.

c)　組織の事業プロセスへの品質マネジメントシス

agement system requirements into the organization's business processes;

d) promoting the use of the process approach and risk-based thinking;

e) ensuring that the resources needed for the quality management system are available;

f) communicating the importance of effective quality management and of conforming to the quality management system requirements;

g) ensuring that the quality management system achieves its intended results;

h) engaging, directing and supporting persons to contribute to the effectiveness of the quality management system;

i) promoting improvement;

j) supporting other relevant management roles to demonstrate their leadership as it applies to their areas of responsibility.

NOTE Reference to "business" in this International Standard can be interpreted broadly to mean those activities that are core to the purposes of the organization's existence, whether the or-

テム要求事項の統合を確実にする.

d)　プロセスアプローチ及びリスクに基づく考え方の利用を促進する.

e)　品質マネジメントシステムに必要な資源が利用可能であることを確実にする.

f)　有効な品質マネジメント及び品質マネジメントシステム要求事項への適合の重要性を伝達する.

g)　品質マネジメントシステムがその意図した結果を達成することを確実にする.

h)　品質マネジメントシステムの有効性に寄与するよう人々を積極的に参加させ，指揮し，支援する.

i)　改善を促進する.

j)　その他の関連する管理層がその責任の領域においてリーダーシップを実証するよう，管理層の役割を支援する.

注記　この規格で“事業”という場合，それは，組織が公的か私的か，営利か非営利かを問わず，組織の存在の目的の中核となる活動という広義の意味で解釈され得る.

ganization is public, private, for profit or not for profit.

5.1.2 Customer focus

Top management shall demonstrate leadership and commitment with respect to customer focus by ensuring that:

a) customer and applicable statutory and regulatory requirements are determined, understood and consistently met;
b) the risks and opportunities that can affect conformity of products and services and the ability to enhance customer satisfaction are determined and addressed;
c) the focus on enhancing customer satisfaction is maintained.

5.2 Policy

5.2.1 Establishing the quality policy

Top management shall establish, implement and maintain a quality policy that:

a) is appropriate to the purpose and context of the organization and supports its strategic direction;

5.1.2　顧客重視

トップマネジメントは，次の事項を確実にすることによって，顧客重視に関するリーダーシップ及びコミットメントを実証しなければならない．

a)　顧客要求事項及び適用される法令・規制要求事項を明確にし，理解し，一貫してそれを満たしている．

b)　製品及びサービスの適合並びに顧客満足を向上させる能力に影響を与え得る，リスク及び機会を決定し，取り組んでいる．

c)　顧客満足向上の重視が維持されている．

5.2　方針

5.2.1　品質方針の確立

トップマネジメントは，次の事項を満たす品質方針を確立し，実施し，維持しなければならない．

a)　組織の目的及び状況に対して適切であり，組織の戦略的な方向性を支援する．

b) provides a framework for setting quality objectives;
c) includes a commitment to satisfy applicable requirements;
d) includes a commitment to continual improvement of the quality management system.

5.2.2 Communicating the quality policy

The quality policy shall:

a) be available and be maintained as documented information;
b) be communicated, understood and applied within the organization;
c) be available to relevant interested parties, as appropriate.

5.3 Organizational roles, responsibilities and authorities

Top management shall ensure that the responsibilities and authorities for relevant roles are assigned, communicated and understood within the organization.

b)　品質目標の設定のための枠組みを与える.

c)　適用される要求事項を満たすことへのコミットメントを含む.

d)　品質マネジメントシステムの継続的改善へのコミットメントを含む.

5.2.2　品質方針の伝達

品質方針は，次に示す事項を満たさなければならない.

a)　文書化した情報として利用可能な状態にされ，維持される.

b)　組織内に伝達され，理解され，適用される.

c)　必要に応じて，密接に関連する利害関係者が入手可能である.

5.3　組織の役割，責任及び権限

トップマネジメントは，関連する役割に対して，責任及び権限が割り当てられ，組織内に伝達され，理解されることを確実にしなければならない.

Top management shall assign the responsibility and authority for:

a) ensuring that the quality management system conforms to the requirements of this International Standard;

b) ensuring that the processes are delivering their intended outputs;

c) reporting on the performance of the quality management system and on opportunities for improvement (see 10.1), in particular to top management;

d) ensuring the promotion of customer focus throughout the organization;

e) ensuring that the integrity of the quality management system is maintained when changes to the quality management system are planned and implemented.

6 Planning

6.1 Actions to address risks and opportunities

6.1.1 When planning for the quality management system, the organization shall consider the issues referred to in 4.1 and the requirements re-

トップマネジメントは，次の事項に対して，責任及び権限を割り当てなければならない．

a) 品質マネジメントシステムが，この規格の要求事項に適合することを確実にする．

b) プロセスが，意図したアウトプットを生み出すことを確実にする．

c) 品質マネジメントシステムのパフォーマンス及び改善（**10.1** 参照）の機会を特にトップマネジメントに報告する．

d) 組織全体にわたって，顧客重視を促進することを確実にする．

e) 品質マネジメントシステムへの変更を計画し，実施する場合には，品質マネジメントシステムを“完全に整っている状態”（integrity）に維持することを確実にする．

6 計画

6.1 リスク及び機会への取組み

6.1.1 品質マネジメントシステムの計画を策定するとき，組織は，**4.1** に規定する課題及び **4.2** に規定する要求事項を考慮し，次の事項のために取り組

ferred to in 4.2 and determine the risks and opportunities that need to be addressed to:

a) give assurance that the quality management system can achieve its intended result(s);
b) enhance desirable effects;
c) prevent, or reduce, undesired effects;
d) achieve improvement.

6.1.2 The organization shall plan:

a) actions to address these risks and opportunities;
b) how to:
    1) integrate and implement the actions into its quality management system processes (see 4.4);
    2) evaluate the effectiveness of these actions.

Actions taken to address risks and opportunities shall be proportionate to the potential impact on the conformity of products and services.

NOTE 1 Options to address risks can include avoiding risk, taking risk in order to pursue an opportunity, eliminating the risk source, chang-

む必要があるリスク及び機会を決定しなければならない.

a) 品質マネジメントシステムが，その意図した結果を達成できるという確信を与える.

b) 望ましい影響を増大する.

c) 望ましくない影響を防止又は低減する.

d) 改善を達成する.

6.1.2 組織は，次の事項を計画しなければならない.

a) 上記によって決定したリスク及び機会への取組み

b) 次の事項を行う方法

  1) その取組みの品質マネジメントシステムプロセスへの統合及び実施（**4.4** 参照）

  2) その取組みの有効性の評価

リスク及び機会への取組みは，製品及びサービスの適合への潜在的な影響と見合ったものでなければならない.

注記 1 リスクへの取組みの選択肢には，リスクを回避すること，ある機会を追求するためにそのリスクを取ること，リス

ing the likelihood or consequences, sharing the risk, or retaining risk by informed decision.

NOTE 2 Opportunities can lead to the adoption of new practices, launching new products, opening new markets, addressing new customers, building partnerships, using new technology and other desirable and viable possibilities to address the organization's or its customers' needs.

6.2 Quality objectives and planning to achieve them

6.2.1 The organization shall establish quality objectives at relevant functions, levels and processes needed for the quality management system.

The quality objectives shall:

a) be consistent with the quality policy;

b) be measurable;

c) take into account applicable requirements;

ク源を除去すること，起こりやすさ若しくは結果を変えること，リスクを共有すること，又は情報に基づいた意思決定によってリスクを保有することが含まれ得る．

注記 2 機会は，新たな慣行の採用，新製品の発売，新市場の開拓，新たな顧客への取組み，パートナーシップの構築，新たな技術の使用，及び組織のニーズ又は顧客のニーズに取り組むためのその他の望ましくかつ実行可能な可能性につながり得る．

6.2 品質目標及びそれを達成するための計画策定

6.2.1 組織は，品質マネジメントシステムに必要な，関連する機能，階層及びプロセスにおいて，品質目標を確立しなければならない．

品質目標は，次の事項を満たさなければならない．

a) 品質方針と整合している．

b) 測定可能である．

c) 適用される要求事項を考慮に入れる．

d) be relevant to conformity of products and services and to enhancement of customer satisfaction;
e) be monitored;
f) be communicated;
g) be updated as appropriate.

The organization shall maintain documented information on the quality objectives.

6.2.2 When planning how to achieve its quality objectives, the organization shall determine:

a) what will be done;
b) what resources will be required;
c) who will be responsible;
d) when it will be completed;
e) how the results will be evaluated.

6.3 Planning of changes

When the organization determines the need for changes to the quality management system, the changes shall be carried out in a planned manner (see 4.4).

d) 製品及びサービスの適合，並びに顧客満足の向上に関連している.

e) 監視する.
f) 伝達する.
g) 必要に応じて，更新する.

組織は，品質目標に関する文書化した情報を維持しなければならない.

6.2.2 組織は，品質目標をどのように達成するかについて計画するとき，次の事項を決定しなければならない.

a) 実施事項
b) 必要な資源
c) 責任者
d) 実施事項の完了時期
e) 結果の評価方法

6.3 変更の計画

組織が品質マネジメントシステムの変更の必要性を決定したとき，その変更は，計画的な方法で行わなければならない（**4.4** 参照）.

The organization shall consider:

a) the purpose of the changes and their potential consequences;
b) the integrity of the quality management system;
c) the availability of resources;
d) the allocation or reallocation of responsibilities and authorities.

7 Support

7.1 Resources

7.1.1 General

The organization shall determine and provide the resources needed for the establishment, implementation, maintenance and continual improvement of the quality management system.

The organization shall consider:

a) the capabilities of, and constraints on, existing internal resources;
b) what needs to be obtained from external providers.

組織は，次の事項を考慮しなければならない．

a) 変更の目的，及びそれによって起こり得る結果

b) 品質マネジメントシステムの“完全に整っている状態”（integrity）

c) 資源の利用可能性

d) 責任及び権限の割当て又は再割当て

7 支援

7.1 資源

7.1.1 一般

組織は，品質マネジメントシステムの確立，実施，維持及び継続的改善に必要な資源を明確にし，提供しなければならない．

組織は，次の事項を考慮しなければならない．

a) 既存の内部資源の実現能力及び制約

b) 外部提供者から取得する必要があるもの

7.1.2 People

The organization shall determine and provide the persons necessary for the effective implementation of its quality management system and for the operation and control of its processes.

7.1.3 Infrastructure

The organization shall determine, provide and maintain the infrastructure necessary for the operation of its processes and to achieve conformity of products and services.

NOTE Infrastructure can include:

a) buildings and associated utilities;
b) equipment, including hardware and software;

c) transportation resources;
d) information and communication technology.

7.1.4 Environment for the operation of processes

The organization shall determine, provide and maintain the environment necessary for the oper-

7.1.2 人々

組織は，品質マネジメントシステムの効果的な実施，並びにそのプロセスの運用及び管理のために必要な人々を明確にし，提供しなければならない．

7.1.3 インフラストラクチャ

組織は，プロセスの運用に必要なインフラストラクチャ，並びに製品及びサービスの適合を達成するために必要なインフラストラクチャを明確にし，提供し，維持しなければならない．

注記 インフラストラクチャには，次の事項が含まれ得る．

a) 建物及び関連するユーティリティ
b) 設備．これにはハードウェア及びソフトウェアを含む．
c) 輸送のための資源
d) 情報通信技術

7.1.4 プロセスの運用に関する環境

組織は，プロセスの運用に必要な環境，並びに製品及びサービスの適合を達成するために必要な環境

ation of its processes and to achieve conformity of products and services.

NOTE A suitable environment can be a combination of human and physical factors, such as:

a) social (e.g. non-discriminatory, calm, non-confrontational);
b) psychological (e.g. stress-reducing, burnout prevention, emotionally protective);
c) physical (e.g. temperature, heat, humidity, light, airflow, hygiene, noise).

These factors can differ substantially depending on the products and services provided.

7.1.5 Monitoring and measuring resources

7.1.5.1 General

The organization shall determine and provide the resources needed to ensure valid and reliable results when monitoring or measuring is used to verify the conformity of products and services to requirements.

The organization shall ensure that the resources

を明確にし，提供し，維持しなければならない．

注記 適切な環境は，次のような人的及び物理的要因の組合せであり得る．

a) 社会的要因（例えば，非差別的，平穏，非対立的）

b) 心理的要因（例えば，ストレス軽減，燃え尽き症候群防止，心のケア）

c) 物理的要因（例えば，気温，熱，湿度，光，気流，衛生状態，騒音）

これらの要因は，提供する製品及びサービスによって，大いに異なり得る．

7.1.5 監視及び測定のための資源

7.1.5.1 一般

要求事項に対する製品及びサービスの適合を検証するために監視又は測定を用いる場合，組織は，結果が妥当で信頼できるものであることを確実にするために必要な資源を明確にし，提供しなければならない．

組織は，用意した資源が次の事項を満たすことを

provided:

a) are suitable for the specific type of monitoring and measurement activities being undertaken;
b) are maintained to ensure their continuing fitness for their purpose.

The organization shall retain appropriate documented information as evidence of fitness for purpose of the monitoring and measurement resources.

7.1.5.2 Measurement traceability

When measurement traceability is a requirement, or is considered by the organization to be an essential part of providing confidence in the validity of measurement results, measuring equipment shall be:

a) calibrated or verified, or both, at specified intervals, or prior to use, against measurement standards traceable to international or national measurement standards; when no such standards exist, the basis used for calibration or verification shall be retained as documented information;

確実にしなければならない．

a) 実施する特定の種類の監視及び測定活動に対して適切である．

b) その目的に継続して合致することを確実にするために維持されている．

組織は，監視及び測定のための資源が目的と合致している証拠として，適切な文書化した情報を保持しなければならない．

7.1.5.2 測定のトレーサビリティ

測定のトレーサビリティが要求事項となっている場合，又は組織がそれを測定結果の妥当性に信頼を与えるための不可欠な要素とみなす場合には，測定機器は，次の事項を満たさなければならない．

a) 定められた間隔で又は使用前に，国際計量標準又は国家計量標準に対してトレーサブルである計量標準に照らして校正若しくは検証，又はそれらの両方を行う．そのような標準が存在しない場合には，校正又は検証に用いたよりどころを，文書化した情報として保持する．

b) identified in order to determine their status;

c) safeguarded from adjustments, damage or deterioration that would invalidate the calibration status and subsequent measurement results.

The organization shall determine if the validity of previous measurement results has been adversely affected when measuring equipment is found to be unfit for its intended purpose, and shall take appropriate action as necessary.

7.1.6 Organizational knowledge

The organization shall determine the knowledge necessary for the operation of its processes and to achieve conformity of products and services.

This knowledge shall be maintained and be made available to the extent necessary.

When addressing changing needs and trends, the organization shall consider its current knowledge and determine how to acquire or access any necessary additional knowledge and required up-

b) それらの状態を明確にするために識別を行う.
c) 校正の状態及びそれ以降の測定結果が無効になってしまうような調整，損傷又は劣化から保護する.

測定機器が意図した目的に適していないことが判明した場合，組織は，それまでに測定した結果の妥当性を損なうものであるか否かを明確にし，必要に応じて，適切な処置をとらなければならない.

7.1.6 組織の知識

組織は，プロセスの運用に必要な知識，並びに製品及びサービスの適合を達成するために必要な知識を明確にしなければならない.

この知識を維持し，必要な範囲で利用できる状態にしなければならない.

変化するニーズ及び傾向に取り組む場合，組織は，現在の知識を考慮し，必要な追加の知識及び要求される更新情報を得る方法又はそれらにアクセスする方法を決定しなければならない.

dates.

NOTE 1 Organizational knowledge is knowledge specific to the organization; it is generally gained by experience. It is information that is used and shared to achieve the organization's objectives.

NOTE 2 Organizational knowledge can be based on:

a) internal sources (e.g. intellectual property; knowledge gained from experience; lessons learned from failures and successful projects; capturing and sharing undocumented knowledge and experience; the results of improvements in processes, products and services);
b) external sources (e.g. standards; academia; conferences; gathering knowledge from customers or external providers).

7.2 Competence

The organization shall:

a) determine the necessary competence of person(s) doing work under its control that affects the performance and effectiveness of

注記 1 組織の知識は，組織に固有な知識であり，それは一般的に経験によって得られる．それは，組織の目標を達成するために使用し，共有する情報である．

注記 2 組織の知識は，次の事項に基づいたものであり得る．

a) 内部の知識源（例えば，知的財産，経験から得た知識，成功プロジェクト及び失敗から学んだ教訓，文書化していない知識及び経験の取得及び共有，プロセス，製品及びサービスにおける改善の結果）

b) 外部の知識源（例えば，標準，学界，会議，顧客又は外部の提供者からの知識収集）

7.2 力量

組織は，次の事項を行わなければならない．

a) 品質マネジメントシステムのパフォーマンス及び有効性に影響を与える業務をその管理下で行う人（又は人々）に必要な力量を明確にする．

the quality management system;

b) ensure that these persons are competent on the basis of appropriate education, training, or experience;

c) where applicable, take actions to acquire the necessary competence, and evaluate the effectiveness of the actions taken;

d) retain appropriate documented information as evidence of competence.

NOTE Applicable actions can include, for example, the provision of training to, the mentoring of, or the reassignment of currently employed persons; or the hiring or contracting of competent persons.

7.3 Awareness

The organization shall ensure that persons doing work under the organization's control are aware of:

a) the quality policy;

b) relevant quality objectives;

c) their contribution to the effectiveness of the quality management system, including the

b)　適切な教育，訓練又は経験に基づいて，それらの人々が力量を備えていることを確実にする．

c)　該当する場合には，必ず，必要な力量を身に付けるための処置をとり，とった処置の有効性を評価する．

d)　力量の証拠として，適切な文書化した情報を保持する．

注記　適用される処置には，例えば，現在雇用している人々に対する，教育訓練の提供，指導の実施，配置転換の実施などがあり，また，力量を備えた人々の雇用，そうした人々との契約締結などもあり得る．

7.3　認識

組織は，組織の管理下で働く人々が，次の事項に関して認識をもつことを確実にしなければならない．

a)　品質方針

b)　関連する品質目標

c)　パフォーマンスの向上によって得られる便益を含む，品質マネジメントシステムの有効性に対

benefits of improved performance;

d) the implications of not conforming with the quality management system requirements.

7.4 Communication

The organization shall determine the internal and external communications relevant to the quality management system, including:

a) on what it will communicate;

b) when to communicate;

c) with whom to communicate;

d) how to communicate;

e) who communicates.

7.5 Documented information

7.5.1 General

The organization's quality management system shall include:

a) documented information required by this International Standard;

b) documented information determined by the organization as being necessary for the effectiveness of the quality management system.

する自らの貢献

d) 品質マネジメントシステム要求事項に適合しないことの意味

7.4 コミュニケーション

組織は，次の事項を含む，品質マネジメントシステムに関連する内部及び外部のコミュニケーションを決定しなければならない．

a) コミュニケーションの内容

b) コミュニケーションの実施時期

c) コミュニケーションの対象者

d) コミュニケーションの方法

e) コミュニケーションを行う人

7.5 文書化した情報

7.5.1 一般

組織の品質マネジメントシステムは，次の事項を含まなければならない．

a) この規格が要求する文書化した情報

b) 品質マネジメントシステムの有効性のために必要であると組織が決定した，文書化した情報

NOTE The extent of documented information for a quality management system can differ from one organization to another due to:

— the size of organization and its type of activities, processes, products and services;
— the complexity of processes and their interactions;
— the competence of persons.

7.5.2 Creating and updating

When creating and updating documented information, the organization shall ensure appropriate:

a) identification and description (e.g. a title, date, author, or reference number);
b) format (e.g. language, software version, graphics) and media (e.g. paper, electronic);
c) review and approval for suitability and adequacy.

7.5.3 Control of documented information

7.5.3.1 Documented information required by the quality management system and by this Interna-

注記 品質マネジメントシステムのための文書化した情報の程度は，次のような理由によって，それぞれの組織で異なる場合がある．

— 組織の規模，並びに活動，プロセス，製品及びサービスの種類

— プロセス及びその相互作用の複雑さ

— 人々の力量

7.5.2 作成及び更新

文書化した情報を作成及び更新する際，組織は，次の事項を確実にしなければならない．

a) 適切な識別及び記述（例えば，タイトル，日付，作成者，参照番号）

b) 適切な形式（例えば，言語，ソフトウェアの版，図表）及び媒体（例えば，紙，電子媒体）

c) 適切性及び妥当性に関する，適切なレビュー及び承認

7.5.3 文書化した情報の管理

7.5.3.1 品質マネジメントシステム及びこの規格で要求されている文書化した情報は，次の事項を確

tional Standard shall be controlled to ensure:

a) it is available and suitable for use, where and when it is needed;

b) it is adequately protected (e.g. from loss of confidentiality, improper use, or loss of integrity).

7.5.3.2 For the control of documented information, the organization shall address the following activities, as applicable:

a) distribution, access, retrieval and use;

b) storage and preservation, including preservation of legibility;

c) control of changes (e.g. version control);

d) retention and disposition.

Documented information of external origin determined by the organization to be necessary for the planning and operation of the quality management system shall be identified as appropriate, and be controlled.

Documented information retained as evidence of conformity shall be protected from unintended al-

実にするために，管理しなければならない．

a) 文書化した情報が，必要なときに，必要なところで，入手可能かつ利用に適した状態である．

b) 文書化した情報が十分に保護されている（例えば，機密性の喪失，不適切な使用及び完全性の喪失からの保護）．

7.5.3.2 文書化した情報の管理に当たって，組織は，該当する場合には，必ず，次の行動に取り組まなければならない．

a) 配付，アクセス，検索及び利用

b) 読みやすさが保たれることを含む，保管及び保存

c) 変更の管理（例えば，版の管理）

d) 保持及び廃棄

品質マネジメントシステムの計画及び運用のために組織が必要と決定した外部からの文書化した情報は，必要に応じて識別し，管理しなければならない．

適合の証拠として保持する文書化した情報は，意図しない改変から保護しなければならない．

terations.

NOTE Access can imply a decision regarding the permission to view the documented information only, or the permission and authority to view and change the documented information.

8 Operation

8.1 Operational planning and control

The organization shall plan, implement and control the processes (see 4.4) needed to meet the requirements for the provision of products and services, and to implement the actions determined in Clause 6, by:

a) determining the requirements for the products and services;

b) establishing criteria for:

   1) the processes;

   2) the acceptance of products and services;

c) determining the resources needed to achieve conformity to the product and service requirements;

d) implementing control of the processes in accordance with the criteria;

注記　アクセスとは，文書化した情報の閲覧だけの許可に関する決定，又は文書化した情報の閲覧及び変更の許可及び権限に関する決定を意味し得る．

8　運用

8.1　運用の計画及び管理

組織は，次に示す事項の実施によって，製品及びサービスの提供に関する要求事項を満たすため，並びに箇条**6**で決定した取組みを実施するために必要なプロセスを，計画し，実施し，かつ，管理しなければならない（**4.4**参照）．

a)　製品及びサービスに関する要求事項の明確化

b)　次の事項に関する基準の設定

　1)　プロセス

　2)　製品及びサービスの合否判定

c)　製品及びサービスの要求事項への適合を達成するために必要な資源の明確化

d)　b) の基準に従った，プロセスの管理の実施

e) determining, maintaining and retaining documented information to the extent necessary:
    1) to have confidence that the processes have been carried out as planned;
    2) to demonstrate the conformity of products and services to their requirements.

The output of this planning shall be suitable for the organization's operations.

The organization shall control planned changes and review the consequences of unintended changes, taking action to mitigate any adverse effects, as necessary.

The organization shall ensure that outsourced processes are controlled (see 8.4).

8.2 Requirements for products and services

8.2.1 Customer communication

Communication with customers shall include:

a) providing information relating to products

e) 次の目的のために必要な程度の，文書化した情報の明確化，維持及び保持
 1) プロセスが計画どおりに実施されたという確信をもつ．
 2) 製品及びサービスの要求事項への適合を実証する．

この計画のアウトプットは，組織の運用に適したものでなければならない．

組織は，計画した変更を管理し，意図しない変更によって生じた結果をレビューし，必要に応じて，有害な影響を軽減する処置をとらなければならない．

組織は，外部委託したプロセスが管理されていることを確実にしなければならない（**8.4** 参照）．

8.2 製品及びサービスに関する要求事項

8.2.1 顧客とのコミュニケーション

顧客とのコミュニケーションには，次の事項を含めなければならない．

a) 製品及びサービスに関する情報の提供

and services;

b) handling enquiries, contracts or orders, including changes;

c) obtaining customer feedback relating to products and services, including customer complaints;

d) handling or controlling customer property;

e) establishing specific requirements for contingency actions, when relevant.

8.2.2 Determining the requirements for products and services

When determining the requirements for the products and services to be offered to customers, the organization shall ensure that:

a) the requirements for the products and services are defined, including:

   1) any applicable statutory and regulatory requirements;

   2) those considered necessary by the organization;

b) the organization can meet the claims for the products and services it offers.

b) 引合い，契約又は注文の処理．これらの変更を含む．
c) 苦情を含む，製品及びサービスに関する顧客からのフィードバックの取得
d) 顧客の所有物の取扱い又は管理
e) 関連する場合には，不測の事態への対応に関する特定の要求事項の確立

8.2.2 製品及びサービスに関する要求事項の明確化

顧客に提供する製品及びサービスに関する要求事項を明確にするとき，組織は，次の事項を確実にしなければならない．

a) 次の事項を含む，製品及びサービスの要求事項が定められている．
   1) 適用される法令・規制要求事項
   2) 組織が必要とみなすもの
b) 組織が，提供する製品及びサービスに関して主張していることを満たすことができる．

8.2.3 Review of the requirements for products and services

8.2.3.1 The organization shall ensure that it has the ability to meet the requirements for products and services to be offered to customers. The organization shall conduct a review before committing to supply products and services to a customer, to include:

a) requirements specified by the customer, including the requirements for delivery and postdelivery activities;

b) requirements not stated by the customer, but necessary for the specified or intended use, when known;

c) requirements specified by the organization;

d) statutory and regulatory requirements applicable to the products and services;

e) contract or order requirements differing from those previously expressed.

The organization shall ensure that contract or order requirements differing from those previously defined are resolved.

8.2.3 製品及びサービスに関する要求事項のレビュー

8.2.3.1 組織は，顧客に提供する製品及びサービスに関する要求事項を満たす能力をもつことを確実にしなければならない．組織は，製品及びサービスを顧客に提供することをコミットメントする前に，次の事項を含め，レビューを行わなければならない．

a) 顧客が規定した要求事項．これには引渡し及び引渡し後の活動に関する要求事項を含む．

b) 顧客が明示してはいないが，指定された用途又は意図された用途が既知である場合，それらの用途に応じた要求事項

c) 組織が規定した要求事項

d) 製品及びサービスに適用される法令・規制要求事項

e) 以前に提示されたものと異なる，契約又は注文の要求事項

組織は，契約又は注文の要求事項が以前に定めたものと異なる場合には，それが解決されていることを確実にしなければならない．

The customer's requirements shall be confirmed by the organization before acceptance, when the customer does not provide a documented statement of their requirements.

NOTE In some situations, such as internet sales, a formal review is impractical for each order. Instead, the review can cover relevant product information, such as catalogues.

8.2.3.2 The organization shall retain documented information, as applicable:

a) on the results of the review;
b) on any new requirements for the products and services.

8.2.4 Changes to requirements for products and services

The organization shall ensure that relevant documented information is amended, and that relevant persons are made aware of the changed requirements, when the requirements for products

顧客がその要求事項を書面で示さない場合には，組織は，顧客要求事項を受諾する前に確認しなければならない．

注記 インターネット販売などの幾つかの状況では，注文ごとの正式なレビューは実用的ではない．その代わりとして，レビューには，カタログなどの，関連する製品情報が含まれ得る．

8.2.3.2 組織は，該当する場合には，必ず，次の事項に関する文書化した情報を保持しなければならない．

a) レビューの結果

b) 製品及びサービスに関する新たな要求事項

8.2.4 製品及びサービスに関する要求事項の変更

製品及びサービスに関する要求事項が変更されたときには，組織は，関連する文書化した情報を変更することを確実にしなければならない．また，変更後の要求事項が，関連する人々に理解されていることこ

and services are changed.

8.3 Design and development of products and services

8.3.1 General

The organization shall establish, implement and maintain a design and development process that is appropriate to ensure the subsequent provision of products and services.

8.3.2 Design and development planning

In determining the stages and controls for design and development, the organization shall consider:

a) the nature, duration and complexity of the design and development activities;

b) the required process stages, including applicable design and development reviews;

c) the required design and development verification and validation activities;

d) the responsibilities and authorities involved in the design and development process;

e) the internal and external resource needs for the design and development of products and services;

とを確実にしなければならない.

8.3　製品及びサービスの設計・開発

8.3.1　一般

組織は，以降の製品及びサービスの提供を確実にするために適切な設計・開発プロセスを確立し，実施し，維持しなければならない.

8.3.2　設計・開発の計画

設計・開発の段階及び管理を決定するに当たって，組織は，次の事項を考慮しなければならない.

a)　設計・開発活動の性質，期間及び複雑さ

b)　要求されるプロセス段階．これには適用される設計・開発のレビューを含む.

c)　要求される，設計・開発の検証及び妥当性確認活動

d)　設計・開発プロセスに関する責任及び権限

e)　製品及びサービスの設計・開発のための内部資源及び外部資源の必要性

f) the need to control interfaces between persons involved in the design and development process;
g) the need for involvement of customers and users in the design and development process;
h) the requirements for subsequent provision of products and services;
i) the level of control expected for the design and development process by customers and other relevant interested parties;
j) the documented information needed to demonstrate that design and development requirements have been met.

8.3.3 Design and development inputs

The organization shall determine the requirements essential for the specific types of products and services to be designed and developed. The organization shall consider:

a) functional and performance requirements;
b) information derived from previous similar design and development activities;
c) statutory and regulatory requirements;
d) standards or codes of practice that the orga-

f) 設計・開発プロセスに関与する人々の間のインタフェースの管理の必要性

g) 設計・開発プロセスへの顧客及びユーザの参画の必要性

h) 以降の製品及びサービスの提供に関する要求事項

i) 顧客及びその他の密接に関連する利害関係者によって期待される，設計・開発プロセスの管理レベル

j) 設計・開発の要求事項を満たしていることを実証するために必要な文書化した情報

8.3.3 設計・開発へのインプット

組織は，設計・開発する特定の種類の製品及びサービスに不可欠な要求事項を明確にしなければならない．組織は，次の事項を考慮しなければならない．

a) 機能及びパフォーマンスに関する要求事項

b) 以前の類似の設計・開発活動から得られた情報

c) 法令・規制要求事項

d) 組織が実施することをコミットメントしてい

nization has committed to implement;

e) potential consequences of failure due to the nature of the products and services.

Inputs shall be adequate for design and development purposes, complete and unambiguous.

Conflicting design and development inputs shall be resolved.

The organization shall retain documented information on design and development inputs.

8.3.4 Design and development controls

The organization shall apply controls to the design and development process to ensure that:

a) the results to be achieved are defined;

b) reviews are conducted to evaluate the ability of the results of design and development to meet requirements;

c) verification activities are conducted to ensure that the design and development outputs meet the input requirements;

d) validation activities are conducted to ensure

る，標準又は規範（codes of practice）

e) 製品及びサービスの性質に起因する失敗により起こり得る結果

インプットは，設計・開発の目的に対して適切で，漏れがなく，曖昧でないものでなければならない．

設計・開発へのインプット間の相反は，解決しなければならない．

組織は，設計・開発へのインプットに関する文書化した情報を保持しなければならない．

8.3.4 設計・開発の管理

組織は，次の事項を確実にするために，設計・開発プロセスを管理しなければならない．

a) 達成すべき結果を定める．

b) 設計・開発の結果の，要求事項を満たす能力を評価するために，レビューを行う．

c) 設計・開発からのアウトプットが，インプットの要求事項を満たすことを確実にするために，検証活動を行う．

d) 結果として得られる製品及びサービスが，指定

that the resulting products and services meet the requirements for the specified application or intended use;

e) any necessary actions are taken on problems determined during the reviews, or verification and validation activities;

f) documented information of these activities is retained.

NOTE Design and development reviews, verification and validation have distinct purposes. They can be conducted separately or in any combination, as is suitable for the products and services of the organization.

8.3.5 Design and development outputs

The organization shall ensure that design and development outputs:

a) meet the input requirements;

b) are adequate for the subsequent processes for the provision of products and services;

c) include or reference monitoring and measuring requirements, as appropriate, and acceptance criteria;

された用途又は意図された用途に応じた要求事項を満たすことを確実にするために，妥当性確認活動を行う．

e) レビュー，又は検証及び妥当性確認の活動中に明確になった問題に対して必要な処置をとる．

f) これらの活動についての文書化した情報を保持する．

注記　設計・開発のレビュー，検証及び妥当性確認は，異なる目的をもつ．これらは，組織の製品及びサービスに応じた適切な形で，個別に又は組み合わせて行うことができる．

8.3.5　設計・開発からのアウトプット

組織は，設計・開発からのアウトプットが，次のとおりであることを確実にしなければならない．

a) インプットで与えられた要求事項を満たす．

b) 製品及びサービスの提供に関する以降のプロセスに対して適切である．

c) 必要に応じて，監視及び測定の要求事項，並びに合否判定基準を含むか，又はそれらを参照している．

d) specify the characteristics of the products and services that are essential for their intended purpose and their safe and proper provision.

The organization shall retain documented information on design and development outputs.

8.3.6 Design and development changes

The organization shall identify, review and control changes made during, or subsequent to, the design and development of products and services, to the extent necessary to ensure that there is no adverse impact on conformity to requirements.

The organization shall retain documented information on:

a) design and development changes;
b) the results of reviews;
c) the authorization of the changes;
d) the actions taken to prevent adverse impacts.

d) 意図した目的並びに安全で適切な使用及び提供に不可欠な，製品及びサービスの特性を規定している．

組織は，設計・開発のアウトプットについて，文書化した情報を保持しなければならない．

8.3.6 設計・開発の変更

組織は，要求事項への適合に悪影響を及ぼさないことを確実にするために必要な程度まで，製品及びサービスの設計・開発の間又はそれ以降に行われた変更を識別し，レビューし，管理しなければならない．

組織は，次の事項に関する文書化した情報を保持しなければならない．

a) 設計・開発の変更
b) レビューの結果
c) 変更の許可
d) 悪影響を防止するための処置

8.4 Control of externally provided processes, products and services

8.4.1 General

The organization shall ensure that externally provided processes, products and services conform to requirements.

The organization shall determine the controls to be applied to externally provided processes, products and services when:

a) products and services from external providers are intended for incorporation into the organization's own products and services;

b) products and services are provided directly to the customer(s) by external providers on behalf of the organization;

c) a process, or part of a process, is provided by an external provider as a result of a decision by the organization.

The organization shall determine and apply criteria for the evaluation, selection, monitoring of performance, and re-evaluation of external providers, based on their ability to provide processes

8.4 外部から提供されるプロセス，製品及びサービスの管理

8.4.1 一般

組織は，外部から提供されるプロセス，製品及びサービスが，要求事項に適合していることを確実にしなければならない．

組織は，次の事項に該当する場合には，外部から提供されるプロセス，製品及びサービスに適用する管理を決定しなければならない．

a) 外部提供者からの製品及びサービスが，組織自身の製品及びサービスに組み込むことを意図したものである場合

b) 製品及びサービスが，組織に代わって，外部提供者から直接顧客に提供される場合

c) プロセス又はプロセスの一部が，組織の決定の結果として，外部提供者から提供される場合

組織は，要求事項に従ってプロセス又は製品・サービスを提供する外部提供者の能力に基づいて，外部提供者の評価，選択，パフォーマンスの監視，及び再評価を行うための基準を決定し，適用しなけれ

or products and services in accordance with requirements. The organization shall retain documented information of these activities and any necessary actions arising from the evaluations.

8.4.2 Type and extent of control

The organization shall ensure that externally provided processes, products and services do not adversely affect the organization's ability to consistently deliver conforming products and services to its customers.

The organization shall:

a) ensure that externally provided processes remain within the control of its quality management system;
b) define both the controls that it intends to apply to an external provider and those it intends to apply to the resulting output;
c) take into consideration:
    1) the potential impact of the externally provided processes, products and services on the organization's ability to consistently meet customer and applicable

ばならない．組織は，これらの活動及びその評価によって生じる必要な処置について，文書化した情報を保持しなければならない．

8.4.2 管理の方式及び程度

組織は，外部から提供されるプロセス，製品及びサービスが，顧客に一貫して適合した製品及びサービスを引き渡す組織の能力に悪影響を及ぼさないことを確実にしなければならない．

組織は，次の事項を行わなければならない．

a) 外部から提供されるプロセスを組織の品質マネジメントシステムの管理下にとどめることを，確実にする．

b) 外部提供者に適用するための管理，及びそのアウトプットに適用するための管理の両方を定める．

c) 次の事項を考慮に入れる．

   1) 外部から提供されるプロセス，製品及びサービスが，顧客要求事項及び適用される法令・規制要求事項を一貫して満たす組織の能力に与える潜在的な影響

statutory and regulatory requirements;

2) the effectiveness of the controls applied by the external provider;

d) determine the verification, or other activities, necessary to ensure that the externally provided processes, products and services meet requirements.

8.4.3 Information for external providers

The organization shall ensure the adequacy of requirements prior to their communication to the external provider.

The organization shall communicate to external providers its requirements for:

a) the processes, products and services to be provided;

b) the approval of:

   1) products and services;

   2) methods, processes and equipment;

   3) the release of products and services;

c) competence, including any required qualification of persons;

d) the external providers' interactions with the

2) 外部提供者によって適用される管理の有効性

d) 外部から提供されるプロセス，製品及びサービスが要求事項を満たすことを確実にするために必要な検証又はその他の活動を明確にする.

8.4.3 外部提供者に対する情報

組織は，外部提供者に伝達する前に，要求事項が妥当であることを確実にしなければならない.

組織は，次の事項に関する要求事項を，外部提供者に伝達しなければならない.

a) 提供されるプロセス，製品及びサービス

b) 次の事項についての承認
   1) 製品及びサービス
   2) 方法，プロセス及び設備
   3) 製品及びサービスのリリース

c) 人々の力量．これには必要な適格性を含む.

d) 組織と外部提供者との相互作用

organization;

e) control and monitoring of the external providers' performance to be applied by the organization;

f) verification or validation activities that the organization, or its customer, intends to perform at the external providers' premises.

8.5 Production and service provision

8.5.1 Control of production and service provision

The organization shall implement production and service provision under controlled conditions.

Controlled conditions shall include, as applicable:

a) the availability of documented information that defines:
   1) the characteristics of the products to be produced, the services to be provided, or the activities to be performed;
   2) the results to be achieved;

b) the availability and use of suitable monitoring and measuring resources;

e) 組織が適用する，外部提供者のパフォーマンスの管理及び監視

f) 組織又はその顧客が外部提供者先での実施を意図している検証又は妥当性確認活動

8.5 製造及びサービス提供

8.5.1 製造及びサービス提供の管理

組織は，製造及びサービス提供を，管理された状態で実行しなければならない．

管理された状態には，次の事項のうち，該当するものについては，必ず，含めなければならない．

a) 次の事項を定めた文書化した情報を利用できるようにする．

  1) 製造する製品，提供するサービス，又は実施する活動の特性．

  2) 達成すべき結果

b) 監視及び測定のための適切な資源を利用できるようにし，かつ，使用する．

c) the implementation of monitoring and measurement activities at appropriate stages to verify that criteria for control of processes or outputs, and acceptance criteria for products and services, have been met;

d) the use of suitable infrastructure and environment for the operation of processes;

e) the appointment of competent persons, including any required qualification;

f) the validation, and periodic revalidation, of the ability to achieve planned results of the processes for production and service provision, where the resulting output cannot be verified by subsequent monitoring or measurement;

g) the implementation of actions to prevent human error;

h) the implementation of release, delivery and post-delivery activities.

8.5.2 Identification and traceability

The organization shall use suitable means to identify outputs when it is necessary to ensure the conformity of products and services.

c)　プロセス又はアウトプットの管理基準，並びに製品及びサービスの合否判定基準を満たしていることを検証するために，適切な段階で監視及び測定活動を実施する．

d)　プロセスの運用のための適切なインフラストラクチャ及び環境を使用する．

e)　必要な適格性を含め，力量を備えた人々を任命する．

f)　製造及びサービス提供のプロセスで結果として生じるアウトプットを，それ以降の監視又は測定で検証することが不可能な場合には，製造及びサービス提供に関するプロセスの，計画した結果を達成する能力について，妥当性確認を行い，定期的に妥当性を再確認する．

g)　ヒューマンエラーを防止するための処置を実施する．

h)　リリース，顧客への引渡し及び引渡し後の活動を実施する．

8.5.2　識別及びトレーサビリティ

製品及びサービスの適合を確実にするために必要な場合，組織は，アウトプットを識別するために，適切な手段を用いなければならない．

The organization shall identify the status of outputs with respect to monitoring and measurement requirements throughout production and service provision.

The organization shall control the unique identification of the outputs when traceability is a requirement, and shall retain the documented information necessary to enable traceability.

8.5.3 Property belonging to customers or external providers

The organization shall exercise care with property belonging to customers or external providers while it is under the organization's control or being used by the organization.

The organization shall identify, verify, protect and safeguard customers' or external providers' property provided for use or incorporation into the products and services.

When the property of a customer or external provider is lost, damaged or otherwise found to be

組織は，製造及びサービス提供の全過程において，監視及び測定の要求事項に関連して，アウトプットの状態を識別しなければならない．

トレーサビリティが要求事項となっている場合には，組織は，アウトプットについて一意の識別を管理し，トレーサビリティを可能とするために必要な文書化した情報を保持しなければならない．

8.5.3 顧客又は外部提供者の所有物

組織は，顧客又は外部提供者の所有物について，それが組織の管理下にある間，又は組織がそれを使用している間は，注意を払わなければならない．

組織は，使用するため又は製品及びサービスに組み込むために提供された顧客又は外部提供者の所有物の識別，検証及び保護・防護を実施しなければならない．

顧客若しくは外部提供者の所有物を紛失若しくは損傷した場合，又はその他これらが使用に適さない

unsuitable for use, the organization shall report this to the customer or external provider and retain documented information on what has occurred.

NOTE A customer's or external provider's property can include materials, components, tools and equipment, premises, intellectual property and personal data.

8.5.4 Preservation

The organization shall preserve the outputs during production and service provision, to the extent necessary to ensure conformity to requirements.

NOTE Preservation can include identification, handling, contamination control, packaging, storage, transmission or transportation, and protection.

8.5.5 Post-delivery activities

The organization shall meet requirements for post-delivery activities associated with the prod-

と判明した場合には，組織は，その旨を顧客又は外部提供者に報告し，発生した事柄について文書化した情報を保持しなければならない．

注記 顧客又は外部提供者の所有物には，材料，部品，道具，設備，施設，知的財産，個人情報などが含まれ得る．

8.5.4 保存

組織は，製造及びサービス提供を行う間，要求事項への適合を確実にするために必要な程度に，アウトプットを保存しなければならない．

注記 保存に関わる考慮事項には，識別，取扱い，汚染防止，包装，保管，伝送又は輸送，及び保護が含まれ得る．

8.5.5 引渡し後の活動

組織は，製品及びサービスに関連する引渡し後の活動に関する要求事項を満たさなければならない．

ucts and services.

In determining the extent of post-delivery activities that are required, the organization shall consider:

a) statutory and regulatory requirements;
b) the potential undesired consequences associated with its products and services;
c) the nature, use and intended lifetime of its products and services;
d) customer requirements;
e) customer feedback.

NOTE Post-delivery activities can include actions under warranty provisions, contractual obligations such as maintenance services, and supplementary services such as recycling or final disposal.

8.5.6 Control of changes

The organization shall review and control changes for production or service provision, to the extent necessary to ensure continuing conformity with requirements.

要求される引渡し後の活動の程度を決定するに当たって，組織は，次の事項を考慮しなければならない．

a) 法令・規制要求事項
b) 製品及びサービスに関連して起こり得る望ましくない結果
c) 製品及びサービスの性質，用途及び意図した耐用期間
d) 顧客要求事項
e) 顧客からのフィードバック

注記 引渡し後の活動には，補償条項（warranty provisions），メンテナンスサービスのような契約義務，及びリサイクル又は最終廃棄のような付帯サービスの下での活動が含まれ得る．

8.5.6 変更の管理

組織は，製造又はサービス提供に関する変更を，要求事項への継続的な適合を確実にするために必要な程度まで，レビューし，管理しなければならない．

The organization shall retain documented information describing the results of the review of changes, the person(s) authorizing the change, and any necessary actions arising from the review.

8.6 Release of products and services

The organization shall implement planned arrangements, at appropriate stages, to verify that the product and service requirements have been met.

The release of products and services to the customer shall not proceed until the planned arrangements have been satisfactorily completed, unless otherwise approved by a relevant authority and, as applicable, by the customer.

The organization shall retain documented information on the release of products and services. The documented information shall include:

a) evidence of conformity with the acceptance criteria;
b) traceability to the person(s) authorizing the

組織は，変更のレビューの結果，変更を正式に許可した人（又は人々）及びレビューから生じた必要な処置を記載した，文書化した情報を保持しなければならない．

8.6 製品及びサービスのリリース

組織は，製品及びサービスの要求事項を満たしていることを検証するために，適切な段階において，計画した取決めを実施しなければならない．

計画した取決めが問題なく完了するまでは，顧客への製品及びサービスのリリースを行ってはならない．ただし，当該の権限をもつ者が承認し，かつ，顧客が承認したとき（該当する場合には，必ず）は，この限りではない．

組織は，製品及びサービスのリリースについて文書化した情報を保持しなければならない．これには，次の事項を含まなければならない．

a) 合否判定基準への適合の証拠

b) リリースを正式に許可した人（又は人々）に対

release.

8.7 Control of nonconforming outputs

8.7.1 The organization shall ensure that outputs that do not conform to their requirements are identified and controlled to prevent their unintended use or delivery.

The organization shall take appropriate action based on the nature of the nonconformity and its effect on the conformity of products and services. This shall also apply to nonconforming products and services detected after delivery of products, during or after the provision of services.

The organization shall deal with nonconforming outputs in one or more of the following ways:

a) correction;

b) segregation, containment, return or suspension of provision of products and services;

c) informing the customer;

d) obtaining authorization for acceptance under concession.

するトレーサビリティ

8.7 不適合なアウトプットの管理

8.7.1 組織は，要求事項に適合しないアウトプットが誤って使用されること又は引き渡されることを防ぐために，それらを識別し，管理することを確実にしなければならない．

組織は，不適合の性質，並びにそれが製品及びサービスの適合に与える影響に基づいて，適切な処置をとらなければならない．これは，製品の引渡し後，サービスの提供中又は提供後に検出された，不適合な製品及びサービスにも適用されなければならない．

組織は，次の一つ以上の方法で，不適合なアウトプットを処理しなければならない．

a) 修正

b) 製品及びサービスの分離，散逸防止，返却又は提供停止

c) 顧客への通知

d) 特別採用による受入の正式な許可の取得

Conformity to the requirements shall be verified when nonconforming outputs are corrected.

8.7.2 The organization shall retain documented information that:

a) describes the nonconformity;

b) describes the actions taken;

c) describes any concessions obtained;

d) identifies the authority deciding the action in respect of the nonconformity.

9 Performance evaluation

9.1 Monitoring, measurement, analysis and evaluation

9.1.1 General

The organization shall determine:

a) what needs to be monitored and measured;

b) the methods for monitoring, measurement, analysis and evaluation needed to ensure valid results;

c) when the monitoring and measuring shall be performed;

d) when the results from monitoring and measurement shall be analysed and evaluated.

不適合なアウトプットに修正を施したときには，要求事項への適合を検証しなければならない.

8.7.2　組織は，次の事項を満たす文書化した情報を保持しなければならない.

a)　不適合が記載されている.

b)　とった処置が記載されている.

c)　取得した特別採用が記載されている.

d)　不適合に関する処置について決定する権限をもつ者を特定している.

9　パフォーマンス評価

9.1　監視，測定，分析及び評価

9.1.1　一般

組織は，次の事項を決定しなければならない.

a)　監視及び測定が必要な対象

b)　妥当な結果を確実にするために必要な，監視，測定，分析及び評価の方法

c)　監視及び測定の実施時期

d)　監視及び測定の結果の，分析及び評価の時期

The organization shall evaluate the performance and the effectiveness of the quality management system.

The organization shall retain appropriate documented information as evidence of the results.

9.1.2 Customer satisfaction

The organization shall monitor customers' perceptions of the degree to which their needs and expectations have been fulfilled. The organization shall determine the methods for obtaining, monitoring and reviewing this information.

NOTE Examples of monitoring customer perceptions can include customer surveys, customer feedback on delivered products and services, meetings with customers, market-share analysis, compliments, warranty claims and dealer reports.

9.1.3 Analysis and evaluation

The organization shall analyse and evaluate appropriate data and information arising from mon-

組織は，品質マネジメントシステムのパフォーマンス及び有効性を評価しなければならない.

組織は，この結果の証拠として，適切な文書化した情報を保持しなければならない.

9.1.2　顧客満足

組織は，顧客のニーズ及び期待が満たされている程度について，顧客がどのように受け止めているかを監視しなければならない. 組織は，この情報の入手，監視及びレビューの方法を決定しなければならない.

注記　顧客の受け止め方の監視には，例えば，顧客調査，提供した製品及びサービスに関する顧客からのフィードバック，顧客との会合，市場シェアの分析，顧客からの賛辞，補償請求及びディーラ報告が含まれ得る.

9.1.3　分析及び評価

組織は，監視及び測定からの適切なデータ及び情報を分析し，評価しなければならない.

itoring and measurement.

The results of analysis shall be used to evaluate:

a) conformity of products and services;
b) the degree of customer satisfaction;
c) the performance and effectiveness of the quality management system;
d) if planning has been implemented effectively;
e) the effectiveness of actions taken to address risks and opportunities;
f) the performance of external providers;
g) the need for improvements to the quality management system.

NOTE Methods to analyse data can include statistical techniques.

9.2 Internal audit

9.2.1 The organization shall conduct internal audits at planned intervals to provide information on whether the quality management system:

a) conforms to:

分析の結果は，次の事項を評価するために用いなければならない．

a)　製品及びサービスの適合
b)　顧客満足度
c)　品質マネジメントシステムのパフォーマンス及び有効性
d)　計画が効果的に実施されたかどうか．
e)　リスク及び機会への取組みの有効性

f)　外部提供者のパフォーマンス
g)　品質マネジメントシステムの改善の必要性

注記　データを分析する方法には，統計的手法が含まれ得る．

9.2　内部監査

9.2.1　組織は，品質マネジメントシステムが次の状況にあるか否かに関する情報を提供するために，あらかじめ定めた間隔で内部監査を実施しなければならない．

a)　次の事項に適合している．

1) the organization's own requirements for its quality management system;
2) the requirements of this International Standard;

b) is effectively implemented and maintained.

9.2.2 The organization shall:

a) plan, establish, implement and maintain an audit programme(s) including the frequency, methods, responsibilities, planning requirements and reporting, which shall take into consideration the importance of the processes concerned, changes affecting the organization, and the results of previous audits;
b) define the audit criteria and scope for each audit;
c) select auditors and conduct audits to ensure objectivity and the impartiality of the audit process;
d) ensure that the results of the audits are reported to relevant management;
e) take appropriate correction and corrective actions without undue delay;

1) 品質マネジメントシステムに関して，組織自体が規定した要求事項

2) この規格の要求事項

b) 有効に実施され，維持されている．

9.2.2 組織は，次に示す事項を行わなければならない．

a) 頻度，方法，責任，計画要求事項及び報告を含む，監査プログラムの計画，確立，実施及び維持．監査プログラムは，関連するプロセスの重要性，組織に影響を及ぼす変更，及び前回までの監査の結果を考慮に入れなければならない．

b) 各監査について，監査基準及び監査範囲を定める．

c) 監査プロセスの客観性及び公平性を確保するために，監査員を選定し，監査を実施する．

d) 監査の結果を関連する管理層に報告することを確実にする．

e) 遅滞なく，適切な修正を行い，是正処置をとる．

f) retain documented information as evidence of the implementation of the audit programme and the audit results.

NOTE See ISO 19011 for guidance.

9.3 Management review

9.3.1 General

Top management shall review the organization's quality management system, at planned intervals, to ensure its continuing suitability, adequacy, effectiveness and alignment with the strategic direction of the organization.

9.3.2 Management review inputs

The management review shall be planned and carried out taking into consideration:

a) the status of actions from previous management reviews;

b) changes in external and internal issues that are relevant to the quality management system;

c) information on the performance and effectiveness of the quality management system,

f)　監査プログラムの実施及び監査結果の証拠として，文書化した情報を保持する．

注記　手引として **JIS Q 19011** を参照．

9.3　マネジメントレビュー

9.3.1　一般

トップマネジメントは，組織の品質マネジメントシステムが，引き続き，適切，妥当かつ有効で更に組織の戦略的な方向性と一致していることを確実にするために，あらかじめ定めた間隔で，品質マネジメントシステムをレビューしなければならない．

9.3.2　マネジメントレビューへのインプット

マネジメントレビューは，次の事項を考慮して計画し，実施しなければならない．

a)　前回までのマネジメントレビューの結果とった処置の状況

b)　品質マネジメントシステムに関連する外部及び内部の課題の変化

c)　次に示す傾向を含めた，品質マネジメントシステムのパフォーマンス及び有効性に関する情報

including trends in:

1) customer satisfaction and feedback from relevant interested parties;
2) the extent to which quality objectives have been met;
3) process performance and conformity of products and services;
4) nonconformities and corrective actions;
5) monitoring and measurement results;
6) audit results;
7) the performance of external providers;

d) the adequacy of resources;
e) the effectiveness of actions taken to address risks and opportunities (see 6.1);
f) opportunities for improvement.

9.3.3 Management review outputs

The outputs of the management review shall include decisions and actions related to:

a) opportunities for improvement;
b) any need for changes to the quality management system;
c) resource needs.

1)　顧客満足及び密接に関連する利害関係者からのフィードバック
2)　品質目標が満たされている程度

3)　プロセスのパフォーマンス，並びに製品及びサービスの適合
4)　不適合及び是正処置
5)　監視及び測定の結果
6)　監査結果
7)　外部提供者のパフォーマンス

d)　資源の妥当性
e)　リスク及び機会への取組みの有効性（**6.1** 参照）

f)　改善の機会

9.3.3　マネジメントレビューからのアウトプット

マネジメントレビューからのアウトプットには，次の事項に関する決定及び処置を含めなければならない．

a)　改善の機会
b)　品質マネジメントシステムのあらゆる変更の必要性
c)　資源の必要性

The organization shall retain documented information as evidence of the results of management reviews.

10 Improvement

10.1 General

The organization shall determine and select opportunities for improvement and implement any necessary actions to meet customer requirements and enhance customer satisfaction.

These shall include:

a) improving products and services to meet requirements as well as to address future needs and expectations;

b) correcting, preventing or reducing undesired effects;

c) improving the performance and effectiveness of the quality management system.

NOTE Examples of improvement can include correction, corrective action, continual improvement, breakthrough change, innovation and re-organization.

組織は，マネジメントレビューの結果の証拠として，文書化した情報を保持しなければならない．

10 改善

10.1 一般

組織は，顧客要求事項を満たし，顧客満足を向上させるために，改善の機会を明確にし，選択しなければならず，また，必要な取組みを実施しなければならない．

これには，次の事項を含めなければならない．

a) 要求事項を満たすため，並びに将来のニーズ及び期待に取り組むための，製品及びサービスの改善

b) 望ましくない影響の修正，防止又は低減

c) 品質マネジメントシステムのパフォーマンス及び有効性の改善

注記 改善には，例えば，修正，是正処置，継続的改善，現状を打破する変更，革新及び組織再編が含まれ得る．

10.2 Nonconformity and corrective action

10.2.1 When a nonconformity occurs, including any arising from complaints, the organization shall:

a) react to the nonconformity and, as applicable:

  1) take action to control and correct it;

  2) deal with the consequences;

b) evaluate the need for action to eliminate the cause(s) of the nonconformity, in order that it does not recur or occur elsewhere, by:

  1) reviewing and analysing the nonconformity;

  2) determining the causes of the nonconformity;

  3) determining if similar nonconformities exist, or could potentially occur;

c) implement any action needed;

d) review the effectiveness of any corrective action taken;

e) update risks and opportunities determined

10.2 不適合及び是正処置

10.2.1 苦情から生じたものを含め，不適合が発生した場合，組織は，次の事項を行わなければならない．

a) その不適合に対処し，該当する場合には，必ず，次の事項を行う．
   1) その不適合を管理し，修正するための処置をとる．
   2) その不適合によって起こった結果に対処する．

b) その不適合が再発又は他のところで発生しないようにするため，次の事項によって，その不適合の原因を除去するための処置をとる必要性を評価する．
   1) その不適合をレビューし，分析する．
   2) その不適合の原因を明確にする．
   3) 類似の不適合の有無，又はそれが発生する可能性を明確にする．

c) 必要な処置を実施する．

d) とった全ての是正処置の有効性をレビューする．

e) 必要な場合には，計画の策定段階で決定したリ

during planning, if necessary;

f) make changes to the quality management system, if necessary.

Corrective actions shall be appropriate to the effects of the nonconformities encountered.

10.2.2 The organization shall retain documented information as evidence of:

a) the nature of the nonconformities and any subsequent actions taken;

b) the results of any corrective action.

10.3 Continual improvement

The organization shall continually improve the suitability, adequacy and effectiveness of the quality management system.

The organization shall consider the results of analysis and evaluation, and the outputs from management review, to determine if there are needs or opportunities that shall be addressed as part of continual improvement.

スク及び機会を更新する.

f) 必要な場合には，品質マネジメントシステムの変更を行う.

是正処置は，検出された不適合のもつ影響に応じたものでなければならない.

10.2.2 組織は，次に示す事項の証拠として，文書化した情報を保持しなければならない.

a) 不適合の性質及びそれに対してとったあらゆる処置

b) 是正処置の結果

10.3 継続的改善

組織は，品質マネジメントシステムの適切性，妥当性及び有効性を継続的に改善しなければならない.

組織は，継続的改善の一環として取り組まなければならない必要性又は機会があるかどうかを明確にするために，分析及び評価の結果並びにマネジメントレビューからのアウトプットを検討しなければならない.

Annex A
(informative)

Clarification of new structure, terminology and concepts

A.1 Structure and terminology

The clause structure (i.e. clause sequence) and some of the terminology of this edition of this International Standard, in comparison with the previous edition (ISO 9001:2008), have been changed to improve alignment with other management systems standards.

There is no requirement in this International Standard for its structure and terminology to be applied to the documented information of an organization's quality management system.

The structure of clauses is intended to provide a coherent presentation of requirements, rather than a model for documenting an organization's policies, objectives and processes. The structure and content of documented information related to a quality management system can often be more

附属書 A
（参考）

新たな構造，用語及び概念の明確化

A.1　構造及び用語

この規格の箇条の構造（すなわち，箇条の順序）及び一部の用語は，他のマネジメントシステム規格との一致性を向上させるために，旧規格である **JIS Q 9001**:2008 から変更している．

この規格では，組織の品質マネジメントシステムの文書化した情報にこの規格の構造及び用語を適用することは要求していない．

箇条の構造は，組織の方針，目標及びプロセスを文書化するためのモデルを示すというよりも，要求事項を首尾一貫した形で示すことを意図している．品質マネジメントシステムに関係する，文書化した情報の構造及び内容は，その情報が組織によって運用されるプロセスと他の目的のために維持される情

relevant to its users if it relates to both the processes operated by the organization and information maintained for other purposes.

There is no requirement for the terms used by an organization to be replaced by the terms used in this International Standard to specify quality management system requirements. Organizations can choose to use terms which suit their operations (e.g. using "records", "documentation" or "protocols" rather than "documented information"; or "supplier", "partner" or "vendor" rather than "external provider"). Table A.1 shows the major differences in terminology between this edition of this International Standard and the previous edition.

A.2 Products and services

ISO 9001:2008 used the term "product" to include all output categories. This edition of this International Standard uses "products and services". "Products and services" include all output categories (hardware, services, software and processed materials).

報との両方に関係する場合は，より密接に利用者に関連するものになることが多い.

組織で用いる用語を，品質マネジメントシステム要求事項を規定するためにこの規格で用いている用語に置き換えることは要求していない．組織は，それぞれの運用に適した用語を用いることを選択できる（例えば，“文書化した情報”ではなく，“記録”，“文書類”又は“プロトコル”を用いる．“外部提供者”ではなく，“供給者”，“パートナ”又は“販売者”を用いる.）．**表A.1**に，この規格と**JIS Q 9001**:2008との間の用語における主な相違点を示す.

A.2　製品及びサービス

JIS Q 9001:2008では，アウトプットの全ての分類を含めるために，“製品”という用語を用いたが，この規格では，“製品及びサービス”を用いている．“製品及びサービス”は，アウトプットの全ての分類（ハードウェア，サービス，ソフトウェア及び素材製品）を含んでいる.

Table A.1 — Major differences in terminology between ISO 9001:2008 and ISO 9001:2015

ISO 9001:2008	ISO 9001:2015
Products	Products and services
Exclusions	Not used (See Clause A.5 for clarification of applicability)
Management representative	Not used (Similar responsibilities and authorities are assigned but no requirement for a single management representative)
Documentation, quality manual, documented procedures, records	Documented information
Work environment	Environment for the operation of processes
Monitoring and measuring equipment	Monitoring and measuring resources
Purchased product	Externally provided products and services
Supplier	External provider

The specific inclusion of "services" is intended to highlight the differences between products and services in the application of some requirements. The characteristic of services is that at least part of the output is realized at the interface with the customer. This means, for example, that conformity to requirements cannot necessarily be confirmed before service delivery.

表 A.1 — JIS Q 9001:2008 とこの規格との間の主な用語の相違点

JIS Q 9001:2008	この規格
製品	製品及びサービス
除外	該当なし （適用可能性の明確化については，**A.5** 参照）
管理責任者	該当なし （類似の責任及び権限は割り当てられているが，一人の管理責任者という要求事項はない.）
文書類，品質マニュアル，文書化された手順，記録	文書化した情報
作業環境	プロセスの運用に関する環境
監視機器及び測定機器	監視及び測定のための資源
購買製品	外部から提供される製品及びサービス
供給者	外部提供者

特に“サービス”を含めたのは，幾つかの要求事項の適用において，製品とサービスとの間の違いを強調するためである．サービスの特性とは，少なくともアウトプットの一部が，顧客とのインタフェースで実現されることである．これは，例えば，要求事項への適合がサービスの提供前に確認できるとは限らないことを意味している．

In most cases, products and services are used together. Most outputs that organizations provide to customers, or are supplied to them by external providers, include both products and services. For example, a tangible or intangible product can have some associated service or a service can have some associated tangible or intangible product.

A.3 Understanding the needs and expectations of interested parties

Subclause 4.2 specifies requirements for the organization to determine the interested parties that are relevant to the quality management system and the requirements of those interested parties. However, 4.2 does not imply extension of quality management system requirements beyond the scope of this International Standard. As stated in the scope, this International Standard is applicable where an organization needs to demonstrate its ability to consistently provide products and services that meet customer and applicable statutory and regulatory requirements, and aims to enhance customer satisfaction.

多くの場合，“製品”及び“サービス”は，一緒に用いられている．組織が顧客に提供する，又は外部提供者から組織に供給される多くのアウトプットは，製品とサービスの両方を含んでいる．例えば，有形若しくは無形の製品が関連するサービスを伴っている場合，又はサービスが関連する有形若しくは無形の製品を伴っている場合がある．

A.3　利害関係者のニーズ及び期待の理解

4.2 は，組織が品質マネジメントシステムに密接に関連する利害関係者，及びそれらの利害関係者の要求事項を明確にするための要求事項を規定している．しかし，**4.2** は，品質マネジメントシステム要求事項が，この規格の適用範囲を越えて拡大されることを意味しているのではない．適用範囲で規定しているように，この規格は，組織が顧客要求事項及び適用される法令・規制要求事項を満たした製品又はサービスを一貫して提供する能力をもつことを実証する必要がある場合，並びに顧客満足の向上を目指す場合に適用できる．

There is no requirement in this International Standard for the organization to consider interested parties where it has decided that those parties are not relevant to its quality management system. It is for the organization to decide if a particular requirement of a relevant interested party is relevant to its quality management system.

A.4 Risk-based thinking

The concept of risk-based thinking has been implicit in previous editions of this International Standard, e.g. through requirements for planning, review and improvement. This International Standard specifies requirements for the organization to understand its context (see 4.1) and determine risks as a basis for planning (see 6.1). This represents the application of risk-based thinking to planning and implementing quality management system processes (see 4.4) and will assist in determining the extent of documented information.

One of the key purposes of a quality management

この規格では，組織に対し，組織が自らの品質マネジメントシステムに密接に関連しないと決定した利害関係者を考慮することは要求していない．密接に関連する利害関係者の特定の要求事項が自らの品質マネジメントシステムに密接に関連するかどうかを決定するのは，組織である．

A.4　リスクに基づく考え方

リスクに基づく考え方の概念は，例えば，計画策定，レビュー及び改善に関する要求事項を通じて，従来からこの規格の旧版に含まれていた．この規格は，組織が自らの状況を理解し（**4.1** 参照），計画策定の基礎としてリスクを決定する（**6.1** 参照）ための要求事項を規定している．これは，リスクに基づく考え方を品質マネジメントシステムプロセスの計画策定及び実施に適用することを示しており（**4.4** 参照），文書化した情報の程度を決定する際に役立つ．

品質マネジメントシステムの主な目的の一つは，

system is to act as a preventive tool. Consequently, this International Standard does not have a separate clause or subclause on preventive action. The concept of preventive action is expressed through the use of risk-based thinking in formulating quality management system requirements.

The risk-based thinking applied in this International Standard has enabled some reduction in prescriptive requirements and their replacement by performance-based requirements. There is greater flexibility than in ISO 9001:2008 in the requirements for processes, documented information and organizational responsibilities.

Although 6.1 specifies that the organization shall plan actions to address risks, there is no requirement for formal methods for risk management or a documented risk management process. Organizations can decide whether or not to develop a more extensive risk management methodology than is required by this International Standard, e.g. through the application of other guidance or standards.

予防ツールとしての役割を果たすことである．したがって，この規格には，予防処置に関する個別の箇条又は細分箇条はない．予防処置の概念は，品質マネジメントシステム要求事項を策定する際に，リスクに基づく考え方を用いることで示されている．

この規格で適用されているリスクに基づく考え方によって，規範的な要求事項の一部削減，及びパフォーマンスに基づく要求事項によるそれらの置換えが可能となった．プロセス，文書化した情報及び組織の責任に関する要求事項の柔軟性は，**JIS Q 9001**:2008 よりも高まっている．

6.1 は，組織がリスクへの取組みを計画しなければならないことを規定しているが，リスクマネジメントのための厳密な方法又は文書化したリスクマネジメントプロセスは要求していない．組織は，例えば，他の手引又は規格の適用を通じて，この規格で要求しているよりも広範なリスクマネジメントの方法論を展開するかどうかを決定することができる．

Not all the processes of a quality management system represent the same level of risk in terms of the organization's ability to meet its objectives, and the effects of uncertainty are not the same for all organizations. Under the requirements of 6.1, the organization is responsible for its application of risk-based thinking and the actions it takes to address risk, including whether or not to retain documented information as evidence of its determination of risks.

A.5 Applicability

This International Standard does not refer to "exclusions" in relation to the applicability of its requirements to the organization's quality management system. However, an organization can review the applicability of requirements due to the size or complexity of the organization, the management model it adopts, the range of the organization's activities and the nature of the risks and opportunities it encounters.

The requirements for applicability are addressed in 4.3, which defines conditions under which an

品質マネジメントシステムの全てのプロセスが，組織の目標を満たす能力の点から同じレベルのリスクを示すとは限らない．また，不確かさがもたらす影響は，全ての組織にとって同じではない．**6.1** の要求事項の下で，組織は，リスクに基づく考え方の適用，及びリスクを決定した証拠として文書化した情報を保持するかどうかを含めた，リスクへの取組みに対して責任を負う．

A.5　適用可能性

この規格は，組織の品質マネジメントシステムへの要求事項の適用可能性に関する“除外”について言及していない．ただし，組織は，組織の規模又は複雑さ，組織が採用するマネジメントモデル，組織の活動の範囲，並びに組織が遭遇するリスク及び機会の性質による要求事項の適用可能性をレビューすることができる．

4.3 は，適用可能性に関する要求事項を規定しており，そこに定める条件に基づいて，組織は，ある

organization can decide that a requirement cannot be applied to any of the processes within the scope of its quality management system. The organization can only decide that a requirement is not applicable if its decision will not result in failure to achieve conformity of products and services.

A.6 Documented information

As part of the alignment with other management system standards, a common clause on "documented information" has been adopted without significant change or addition (see 7.5). Where appropriate, text elsewhere in this International Standard has been aligned with its requirements. Consequently, "documented information" is used for all document requirements.

Where ISO 9001:2008 used specific terminology such as "document" or "documented procedures", "quality manual" or "quality plan", this edition of this International Standard defines requirements to "maintain documented information".

Where ISO 9001:2008 used the term "records" to

要求事項が組織の品質マネジメントシステムの適用範囲内でどのプロセスにも適用できないことを決定することができる．その決定が，製品及びサービスの適合が達成されないという結果を招かない場合に限り，組織は，その要求事項を適用不可能と決定することができる．

A.6　文書化した情報

他のマネジメントシステム規格と一致させることの一環として，“文書化した情報”に関する共通箇条を，重要な変更又は追加なく採用した（**7.5** 参照）．必要に応じて，この規格の他の部分の表記を，この要求事項と整合させた．その結果，全ての文書要求事項に“文書化した情報”を用いている．

JIS Q 9001:2008 は，“文書”，“文書化された手順”，“品質マニュアル”，“品質計画書”などの特定の用語を用いていたが，この規格では，“文書化した情報を維持する”という要求事項として規定している．

JIS Q 9001:2008 は，要求事項への適合の証拠

denote documents needed to provide evidence of conformity with requirements, this is now expressed as a requirement to "retain documented information". The organization is responsible for determining what documented information needs to be retained, the period of time for which it is to be retained and the media to be used for its retention.

A requirement to "maintain" documented information does not exclude the possibility that the organization might also need to "retain" that same documented information for a particular purpose, e.g. to retain previous versions of it.

Where this International Standard refers to "information" rather than "documented information" (e.g. in 4.1: "The organization shall monitor and review the information about these external and internal issues"), there is no requirement that this information is to be documented. In such situations, the organization can decide whether or not it is necessary or appropriate to maintain documented information.

の提示に必要な文書を意味するために“記録”という用語を用いていたが，この規格では，“文書化した情報を保持する”という要求事項として表している．組織は，保持する必要のある文書化した情報，保持する期間及び保持のために用いる媒体を決定する責任を負う．

文書化した情報を“維持する”という要求事項は，組織が，特定の目的のため（例えば，文書化した情報の旧版を保持するため）にも，同じものを“保持する”必要があるかもしれないという可能性を除外していない．

この規格のある箇所は，“文書化した情報”というよりも，“情報”に言及している（例えば，**4.1**には，“組織は，これらの外部及び内部の課題に関する情報を監視し，レビューしなければならない．”とある．）．この情報を文書化しなければならないという要求事項はない．組織は，このような状況下で，文書化した情報を維持することが必要又は適切かどうかを決定することができる．

A.7 Organizational knowledge

In 7.1.6, this International Standard addresses the need to determine and manage the knowledge maintained by the organization, to ensure the operation of its processes and that it can achieve conformity of products and services.

Requirements regarding organizational knowledge were introduced for the purpose of:

a) safeguarding the organization from loss of knowledge, e.g.
   - through staff turnover;
   - failure to capture and share information;

b) encouraging the organization to acquire knowledge, e.g.
   - learning from experience;
   - mentoring;
   - benchmarking.

A.8 Control of externally provided processes, products and services

All forms of externally provided processes, products and services are addressed in 8.4, e.g. whether through:

A.7　組織の知識

7.1.6 では，プロセスの運用を確実にし，製品及びサービスの適合を達成することを確実にするために，組織が維持する知識を明確にし，マネジメントすることの必要性を規定している．

組織の知識に関する要求事項は，次のような目的で導入された．

a)　例えば，次のような理由による知識の喪失から組織を保護する．

　—　スタッフの離職

　—　情報の取得及び共有の失敗

b)　例えば，次のような方法で知識を獲得することを組織に推奨する．

　—　経験から学ぶ．

　—　指導を受ける．

　—　ベンチマークする．

A.8　外部から提供されるプロセス，製品及びサービスの管理

8.4 では，例えば，次のような形態のいずれによるかを問わず，外部から提供されるプロセス，製品及びサービスのあらゆる形態について規定している．

a) purchasing from a supplier;

b) an arrangement with an associate company;

c) outsourcing processes to an external provider.

Outsourcing always has the essential characteristic of a service, since it will have at least one activity necessarily performed at the interface between the provider and the organization.

The controls required for external provision can vary widely depending on the nature of the processes, products and services. The organization can apply risk-based thinking to determine the type and extent of controls appropriate to particular external providers and externally provided processes, products and services.

a)　供給者からの購買

b)　関連会社との取決め

c)　外部提供者への，プロセスの外部委託

外部委託は，提供者と組織との間のインタフェースで必ず実行される，少なくとも一つの活動を伴うため，サービスに不可欠な特性を常にもつ．

外部からの提供に対して必要となる管理は，プロセス，製品及びサービスの性質によって大きく異なり得る．組織は，特定の外部提供者並びに外部から提供されるプロセス，製品及びサービスに対して行う，適切な管理の方式及び程度を決定するために，リスクに基づく考え方を適用することができる．

Annex B

(informative)

Other International Standards on quality management and quality management systems developed by ISO/TC 176

The International Standards described in this annex have been developed by ISO/TC 176 to provide supporting information for organizations that apply this International Standard, and to provide guidance for organizations that choose to progress beyond its requirements. Guidance or requirements contained in the documents listed in this annex do not add to, or modify, the requirements of this International Standard.

Table B.1 shows the relationship between these standards and the relevant clauses of this International Standard.

This annex does not include reference to the sector-specific quality management system standards developed by ISO/TC 176.

附属書 B
（参考）

ISO/TC 176 によって作成された品質マネジメント及び品質マネジメントシステムの他の規格類

この附属書に記載する **ISO** 規格類は，この規格を適用する組織に対して支援情報を提供し，その要求事項を超えて進んでいくことを選択する組織のための手引を提供するため，**ISO/TC 176** が作成した．この附属書に記載した文書に含まれる手引又は要求事項は，この規格の要求事項を追加又は変更するものではない．

表 B.1 に，これらの規格類とこの規格の関連する箇条との関係を示す．

この附属書は，**ISO/TC 176** によって作成された特定分野の品質マネジメントシステム規格への参照は含まない．

This International Standard is one of the three core standards developed by ISO/TC 176.

— ISO 9000 *Quality management systems — Fundamentals and vocabulary* provides an essential background for the proper understanding and implementation of this International Standard. The quality management principles are described in detail in ISO 9000 and have been taken into consideration during the development of this International Standard. These principles are not requirements in themselves, but they form the foundation of the requirements specified by this International Standard. ISO 9000 also defines the terms, definitions and concepts used in this International Standard.

— ISO 9001 (this International Standard) specifies requirements aimed primarily at giving confidence in the products and services provided by an organization and thereby enhancing customer satisfaction. Its proper implementation can also be expected to bring other organizational benefits, such as im-

ISO 9001（以下，この附属書の中では“この規格”という．）は，**ISO/TC 176** によって作成された中核となる三規格のうちの一つである．

— **ISO 9000**，Quality management systems — Fundamentals and vocabulary

この規格を適切に理解し，実施するために不可欠な予備知識を与えている．**ISO 9000** に詳述する品質マネジメントの原則は，この規格の作成においても考慮された．この原則自体は要求事項ではないが，この規格に規定する要求事項の基礎となっている．また，**ISO 9000** は，この規格で用いる用語，定義及び概念を定めている．

注記　この国際規格に基づき，**JIS Q 9000** が制定されている．

— **ISO 9001**，Quality management systems — Requirements

主として，組織が提供する製品及びサービスについての信頼を与え，かつ，それによって顧客満足を向上させることを狙いとした要求事項を規定している．これを適切に実施することによって，内部コミュニケーションの改善，組織

proved internal communication, better understanding and control of the organization's processes.

— ISO 9004 *Managing for the sustained success of an organization — A quality management approach* provides guidance for organizations that choose to progress beyond the requirements of this International Standard, to address a broader range of topics that can lead to improvement of the organization's overall performance. ISO 9004 includes guidance on a self-assessment methodology for an organization to be able to evaluate the level of maturity of its quality management system.

The International Standards outlined below can provide assistance to organizations when they are establishing or seeking to improve their quality management systems, their processes or their activities.

— ISO 10001 *Quality management — Customer satisfaction — Guidelines for codes of conduct*

のプロセスのよりよい理解及び管理などの，組織に対する他の便益も期待できる．

注記 この国際規格に基づき，**JIS Q 9001** が制定されている．

— **ISO 9004**，Managing for the sustained success of an organization — A quality management approach

この規格の要求事項を超えて進んでいくことを選択する組織に対し，組織の全体的なパフォーマンスの改善につながり得る，より広範なテーマに取り組むための手引を提供している．**ISO 9004** は，組織が自らの品質マネジメントシステムの成熟度を評価できるようにするための，自己評価の方法論に関する手引を含んでいる．

注記 この国際規格に基づき，**JIS Q 9004** が制定されている．

次の規格類は，組織が品質マネジメントシステム，プロセス若しくは活動を確立し又はそれらの改善を求める際に，組織を支援し得る．

— **ISO 10001**，Quality management — Customer satisfaction — Guidelines for codes of

for organizations provides guidance to an organization in determining that its customer satisfaction provisions meet customer needs and expectations. Its use can enhance customer confidence in an organization and improve customer understanding of what to expect from an organization, thereby reducing the likelihood of misunderstandings and complaints.

— ISO 10002 *Quality management — Customer satisfaction — Guidelines for complaints handling in organizations* provides guidance on the process of handling complaints by recognizing and addressing the needs and expectations of complainants and resolving any complaints received. ISO 10002 provides an open, effective and easy-to-use complaints process, including training of people. It also provides guidance for small businesses.

— ISO 10003 *Quality management — Customer satisfaction — Guidelines for dispute resolution external to organizations* provides guidance for effective and efficient external dis-

conduct for organizations

組織が，その顧客満足の規定が顧客のニーズ及び期待を満たすことを判断する際の手引を提供している．これによって，組織において顧客の信頼を高め，組織に何を期待できるかに関する顧客の理解を高めることで誤解及び苦情の可能性を低減することが可能になる．

注記 この国際規格に基づき，**JIS Q 10001** が制定されている．

— **ISO 10002**, Quality management — Customer satisfaction — Guidelines for complaints handling in organizations

苦情申出者のニーズ及び期待を認識し，対応し，受け取った苦情を解決するという，苦情対応プロセスについての指針を提供している．この指針は，人々の教育・訓練を含め，公開され，効果的で，利用しやすい苦情受付方法を提供し，また，小規模企業のための指針も提供する．

注記 この国際規格に基づき，**JIS Q 10002** が制定されている．

— **ISO 10003**, Quality management — Customer satisfaction — Guidelines for dispute resolution external to organizations

製品に関連する苦情に対する効果的かつ効率

pute resolution for product-related complaints. Dispute resolution gives an avenue of redress when organizations do not remedy a complaint internally. Most complaints can be resolved successfully within the organization, without adversarial procedures.

— ISO 10004 *Quality management — Customer satisfaction — Guidelines for monitoring and measuring* provides guidelines for actions to enhance customer satisfaction and to determine opportunities for improvement of products, processes and attributes that are valued by customers. Such actions can strengthen customer loyalty and help retain customers.

— ISO 10005 *Quality management systems — Guidelines for quality plans* provides guidance on establishing and using quality plans as a means of relating requirements of the process, product, project or contract, to work methods and practices that support product realization. Benefits of establishing a quality

的な外部における紛争解決のための手引を提供している．組織が苦情を内部的に救済しない場合，外部における紛争解決手続が，救済の道を提供する．多くの苦情は，敵対的な手続を必要とすることなく，組織内で解決される可能性がある．

注記　この国際規格に基づき，**JIS Q 10003** が制定されている．

— **ISO 10004**, Quality management — Customer satisfaction — Guidelines for monitoring and measuring

顧客満足を向上させる処置，並びに顧客によって価値が評価された製品，プロセス，及び付帯事項の改善の機会を明確にする処置についての指針を提供している．このような処置は，顧客のロイヤリティを高めることができ，顧客をつなぎとめるのに役立つ．

— **ISO 10005**, Quality management systems — Guidelines for quality plans

プロセス，製品，プロジェクト又は契約の要求事項を，製品実現を支援する作業方法及び慣行に関連付ける手段としての，品質計画書の作成及び使用についての手引を提供している．品質計画書を作成することの便益は，要求事項が

plan are increased confidence that requirements will be met, that processes are in control and the motivation that this can give to those involved.

— ISO 10006 *Quality management systems — Guidelines for quality management in projects* is applicable to projects from the small to large, from simple to complex, from an individual project to being part of a portfolio of projects. ISO 10006 is to be used by personnel managing projects and who need to ensure that their organization is applying the practices contained in the ISO quality management system standards.

— ISO 10007 *Quality management systems — Guidelines for configuration management* is to assist organizations applying configuration management for the technical and administrative direction over the life cycle of a product. Configuration management can be used

満たされ，プロセスが管理されているという確信を高めること，及びそれによって関係者に意欲を与えられることにある．

— **ISO 10006**, Quality management systems — Guidelines for quality management in projects

この指針は，小規模のものから大規模なもの，単純なものから複雑なもの，単独のプロジェクトからプロジェクトのプログラム又はポートフォリオの一部であるものまで，様々なプロジェクトに適用できる．この指針は，プロジェクトを運営管理し，所属組織が品質マネジメントシステムに関する規格の実践の適用を確実にする立場にある要員が用いることを意図している．

注記　この国際規格に基づき，**JIS Q 10006** が制定されている．

— **ISO 10007**, Quality management systems — Guidelines for configuration management

製品のライフサイクルにわたる技術上及び管理上の方向付けのためにコンフィギュレーション管理を適用している組織を支援するためのものである．コンフィギュレーション管理は，こ

to meet the product identification and traceability requirements specified in this International Standard.

— ISO 10008 *Quality management — Customer satisfaction — Guidelines for business-to-consumer electronic commerce transactions* gives guidance on how organizations can implement an effective and efficient business-to-consumer electronic commerce transaction (B2C ECT) system, and thereby provide a basis for consumers to have increased confidence in B2C ECTs, enhance the ability of organizations to satisfy consumers and help reduce complaints and disputes.

— ISO 10012 *Measurement management systems — Requirements for measurement processes and measuring equipment* provides guidance for the management of measurement processes and metrological confirmation of measuring equipment used to support and demonstrate compliance with metrological requirements. ISO 10012 provides quality management criteria for a measurement management system to ensure metrological

の規格に規定する製品の識別及びトレーサビリティの要求事項を満たすために用いることができる．

— **ISO 10008**, Quality management — Customer satisfaction — Guidelines for business-to-consumer electronic commerce transactions

組織がどのように企業と消費者との間の電子商取引システムを効果的かつ効率的に実施できるかについての手引を提供している．これによって，企業・消費者間電子商取引に対する消費者の信頼を高めるための基礎を提供し，消費者を満足させる組織の能力を強化し，苦情及び紛争を減少させるのに役立つ．

— **ISO 10012**, Measurement management systems — Requirements for measurement processes and measuring equipment

計量要求事項への適合性を支援し，実証するために使用する，測定プロセスの運用管理及び測定機器の計量確認に関する手引を提供している．これは，計測マネジメントシステムにおける計量要求事項を満たすことを確実にするための品質マネジメントの基準を提供している．

注記　この国際規格に基づき，**JIS Q 10012**

requirements are met.

— ISO/TR 10013 *Guidelines for quality management system documentation* provides guidelines for the development and maintenance of the documentation necessary for a quality management system. ISO/TR 10013 can be used to document management systems other than those of the ISO quality management system standards, e.g. environmental management systems and safety management systems.

— ISO 10014 *Quality management — Guidelines for realizing financial and economic benefits* is addressed to top management. It provides guidelines for realizing financial and economic benefits through the application of quality management principles. It facilitates application of management principles and selection of methods and tools that enable the sustainable success of an organization.

— ISO 10015 *Quality management — Guidelines for training* provides guidelines to assist organizations in addressing issues related to training. ISO 10015 can be applied whenever

が制定されている．

— **ISO/TR 10013**, Guidelines for quality management system documentation

品質マネジメントシステムにとって必要な文書類の作成及び維持についての指針を提供している．この指針は，品質マネジメントシステムに関する規格以外のマネジメントシステム，例えば，環境マネジメントシステム及び安全マネジメントシステムの文書化のためにも用いることができる．

— **ISO 10014**, Quality management — Guidelines for realizing financial and economic benefits

トップマネジメントに向けたものである．この指針は，品質マネジメントの原則の適用を通して財務的及び経済的便益を実現することについての指針を提供している．この指針は，品質マネジメントの原則の適用，並びに組織の持続的成功を可能にする方法及びツールの選択を容易にする．

— **ISO 10015**, Quality management — Guidelines for training

教育・訓練に関する課題への取組みにおいて組織を支援するための指針を提供している．こ

guidance is required to interpret references to "education" and "training" within the ISO quality management system standards. Any reference to "training" includes all types of education and training.

— ISO/TR 10017 *Guidance on statistical techniques for ISO 9001:2000* explains statistical techniques which follow from the variability that can be observed in the behaviour and results of processes, even under conditions of apparent stability. Statistical techniques allow better use of available data to assist in decision making, and thereby help to continually improve the quality of products and processes to achieve customer satisfaction.

— ISO 10018 *Quality management — Guidelines on people involvement and competence* provides guidelines which influence people involvement and competence. A quality management system depends on the involvement of competent people and the way that they are introduced and integrated into the organization. It is critical to determine, develop and evaluate the knowledge, skills, behav-

の指針は，品質マネジメントシステムに関する規格における“教育・訓練”の解釈について，手引が必要な場合にいつでも適用することができる．“教育・訓練”には，全ての種類の教育及び訓練が含まれる．

— **ISO/TR 10017**, Guidance on statistical techniques for ISO 9001:2000

　明らかな安定状況にある場合でさえ生じる，プロセスの振舞い及び結果において観察され得る変動を扱うために考え出された統計的手法について説明している．統計的手法は，意思決定の支援のために利用可能なデータをより有効に用いることを可能にし，これによって，顧客満足を達成するための製品及びプロセスの品質の継続的改善に役立つ．

— **ISO 10018**, Quality management − Guidelines on people involvement and competence

　人々の参画及び力量に関わる指針を提供している．品質マネジメントシステムは，力量のある人々の参画，及びこれらの人々が組織に導入され，組み込まれる方法によって決まる．必要とされる知識，技能，行動及び作業環境を明確にし，開発し，評価することが重要である．

iour and work environment required.

— ISO 10019 *Guidelines for the selection of quality management system consultants and use of their services* provides guidance for the selection of quality management system consultants and the use of their services. It gives guidance on the process for evaluating the competence of a quality management system consultant and provides confidence that the organization's needs and expectations for the consultant's services will be met.

— ISO 19011 *Guidelines for auditing management systems* provides guidance on the management of an audit programme, on the planning and conducting of an audit of a management system, as well as on the competence and evaluation of an auditor and an audit team. ISO 19011 is intended to apply to auditors, organizations implementing management systems, and organizations needing to conduct audits of management systems.

— **ISO 10019**, Guidelines for the selection of quality management system consultants and use of their services

品質マネジメントシステムコンサルタントの選定及びそのサービスの利用のための手引を提供している．この指針は，品質マネジメントシステムコンサルタントの力量を評価するためのプロセスに関する手引を示し，また，コンサルタントのサービスに対する組織のニーズ及び期待が満たされるだろうという信頼を与える．

注記　この国際規格に基づき，**JIS Q 10019** が制定されている．

— **ISO 19011**, Guidelines for auditing management systems

監査プログラムの管理，マネジメントシステム監査の計画及び実施，並びに監査員及び監査チームの力量及び評価についての手引を提供している．この指針は，監査員，マネジメントシステムを実施する組織，及びマネジメントシステムの監査の実施が必要な組織に適用することを意図している．

注記　この国際規格に基づき，**JIS Q 19011** が制定されている．

Table B.1 — Relationship between other International Standards on quality management and quality management systems and the clauses of this International Standard

Other International Standard	Clause in this International Standard						
	4	5	6	7	8	9	10
ISO 9000	All	All	All	All	All	All	All
ISO 9004	All	All	All	All	All	All	All
ISO 10001					8.2.2, 8.5.1	9.1.2	
ISO 10002					8.2.1,	9.1.2	10.2.1
ISO 10003						9.1.2	
ISO 10004						9.1.2, 9.1.3	
ISO 10005		5.3,	6.1, 6.2	All	All	9.1	10.2
ISO 10006	All	All	All	All	All	All	All
ISO 10007					8.5.2		
ISO 10008	All	All	All	All	All	All	All
ISO 10012				7.1.5			
ISO/TR 10013				7.5			
ISO 10014	All	All	All	All	All	All	All
ISO 10015				7.2			
ISO/TR 10017			6.1	7.1.5		9.1	
ISO 10018	All	All	All	All	Al	All	All
ISO 10019					8.4		
ISO 19011						9.2	

NOTE "All" indicates that all the subclauses in the specific clause of this International Standard are related to the other International Standard.

表 B.1—この規格の箇条と他の品質マネジメント及び品質マネジメントシステムに関する規格類との関係

他の規格類	この規格の箇条						
	箇条 4	箇条 5	箇条 6	箇条 7	箇条 8	箇条 9	箇条10
ISO 9000	全て	全て	全て	全て	全て	全て	全て
ISO 9004	全て	全て	全て	全て	全て	全て	全て
ISO 10001					8.2.2, 8.5.1	9.1.2	
ISO 10002					8.2.1	9.1.2	10.2.1
ISO 10003						9.1.2	
ISO 10004						9.1.2, 9.1.3	
ISO 10005		5.3	6.1, 6.2	全て	全て	9.1	10.2
ISO 10006	全て	全て	全て	全て	全て	全て	全て
ISO 10007					8.5.2		
ISO 10008	全て	全て	全て	全て	全て	全て	全て
ISO 10012				7.1.5			
ISO/TR 10013				7.5			
ISO 10014	全て	全て	全て	全て	全て	全て	全て
ISO 10015				7.2			
ISO/TR 10017			6.1	7.1.5		9.1	
ISO 10018	全て	全て	全て	全て	全て	全て	全て
ISO 10019					8.4		
ISO 19011						9.2	

注記　“全て”は，この規格の特定の箇条の全ての細分箇条が他の規格類と関係していることを意味する．

Bibliography

[1] ISO 9004, *Managing for the sustained success of an organization — A quality management approach*

[2] ISO 10001, *Quality management — Customer satisfaction — Guidelines for codes of conduct for organizations*

[3] ISO 10002, *Quality management — Customer satisfaction — Guidelines for complaints handling in organizations*

[4] ISO 10003, *Quality management — Customer satisfaction — Guidelines for dispute resolution external to organizations*

参 考 文 献

[1] **JIS Q 9004** 組織の持続的成功のための運営管理—品質マネジメントアプローチ

注記 対応国際規格：**ISO 9004**, Managing for the sustained success of an organization — A quality management approach (IDT)

[2] **JIS Q 10001** 品質マネジメント—顧客満足—組織における行動規範のための指針

注記 対応国際規格：**ISO 10001**, Quality management — Customer satisfaction — Guidelines for codes of conduct for organizations (IDT)

[3] **JIS Q 10002** 品質マネジメント—顧客満足—組織における苦情対応のための指針

注記 対応国際規格：**ISO 10002**, Quality management — Customer satisfaction — Guidelines for complaints handling in organizations (IDT)

[4] **JIS Q 10003** 品質マネジメント—顧客満足—組織の外部における紛争解決のための指針

注記 対応国際規格：**ISO 10003**, Quality management — Customer satisfac-

[5] ISO 10004, *Quality management — Customer satisfaction — Guidelines for monitoring and measuring*

[6] ISO 10005, *Quality management systems — Guidelines for quality plans*

[7] ISO 10006, *Quality management systems — Guidelines for quality management in projects*

[8] ISO 10007, *Quality management systems — Guidelines for configuration management*

[9] ISO 10008, *Quality management — Customer satisfaction — Guidelines for business-to-consumer electronic commerce transactions*

[10] ISO 10012, *Measurement management systems — Requirements for measurement processes and measuring equipment*

tion — Guidelines for dispute resolution external to organizations (IDT)

[5] **ISO 10004**, Quality management — Customer satisfaction — Guidelines for monitoring and measuring

[6] **ISO 10005**, Quality management systems — Guidelines for quality plans

[7] **JIS Q 10006**　品質マネジメントシステム—プロジェクトにおける品質マネジメントの指針

注記　対応国際規格：**ISO 10006**, Quality management systems — Guidelines for quality management in projects (IDT)

[8] **ISO 10007**, Quality management systems — Guidelines for configuration management

[9] **ISO 10008**, Quality management — Customer satisfaction — Guidelines for business-to-consumer electronic commerce transactions

[10] **JIS Q 10012**　計測マネジメントシステム—測定プロセス及び測定機器に関する要求事項

注記　対応国際規格：**ISO 10012**, Measurement management systems — Requirements for measurement pro-

[11] ISO/TR 10013, *Guidelines for quality management system documentation*

[12] ISO 10014, *Quality management — Guidelines for realizing financial and economic benefits*

[13] ISO 10015, *Quality management — Guidelines for training*

[14] ISO/TR 10017, *Guidance on statistical techniques for ISO 9001:2000*

[15] ISO 10018, *Quality management — Guidelines on people involvement and competence*

[16] ISO 10019, *Guidelines for the selection of quality management system consultants and use of their services*

[17] ISO 14001, *Environmental management systems — Requirements with guidance for use*

cesses and measuring equipment (IDT)

[11] **ISO/TR 10013**, Guidelines for quality management system documentation

[12] **ISO 10014**, Quality management — Guidelines for realizing financial and economic benefits

[13] **ISO 10015**, Quality management — Guidelines for training

[14] **ISO/TR 10017**, Guidance on statistical techniques for ISO 9001:2000

[15] **ISO 10018**, Quality management — Guidelines on people involvement and competence

[16] **JIS Q 10019**　品質マネジメントシステムコンサルタントの選定及びそのサービスの利用のための指針

注記　対応国際規格：**ISO 10019**, Guidelines for the selection of quality management system consultants and use of their services (IDT)

[17] **JIS Q 14001**　環境マネジメントシステム—要求事項及び利用の手引

注記　対応国際規格：**ISO 14001**, Environmental management systems —

[18] ISO 19011, *Guidelines for auditing management systems*

[19] ISO 31000, *Risk management — Principles and guidelines*

[20] ISO 37500, *Guidance on outsourcing*

[21] ISO/IEC 90003, *Software engineering — Guidelines for the application of ISO 9001:2008 to computer software*

[22] IEC 60300-1, *Dependability management — Part 1: Guidance for management and application*

[23] IEC 61160, *Design review*

[24] Quality management principles, ISO[1)]

[25] Selection and use of the ISO 9000 family of standards, ISO[1)]

[26] ISO 9001 for Small Businesses — What to do,

Requirements with guidance for use（IDT）

[18] **JIS Q 19011** マネジメントシステム監査のための指針

注記 対応国際規格：**ISO 19011**，Guidelines for auditing management systems（IDT）

[19] **JIS Q 31000** リスクマネジメント―原則及び指針

注記 対応国際規格：**ISO 31000**，Risk management — Principles and guidelines（IDT）

[20] **ISO 37500**，Guidance on outsourcing

[21] **ISO/IEC 90003**，Software engineering — Guidelines for the application of ISO 9001:2008 to computer software

[22] **IEC 60300-1**，Dependability management — Part 1: Guidance for management and application

[23] **IEC 61160**，Design review

[24] Quality management principles, ISO 1)

[25] Selection and use of the ISO 9000 family of standards, ISO 1)

[26] ISO 9001 for Small Businesses — What to

ISO[1)]

[27] Integrated use of management system standards, ISO[1)]

[28] www.iso.org/tc176/sc02/public

[29] www.iso.org/tc176/ISO9001AuditingPractices-Group

1) Available from website: http://www.iso.org.

do, ISO [1)]

[27] Integrated use of management system standards, ISO [1)]

[28] www.iso.org/tc176/sc02/public

[29] www.iso.org/tc176/ISO9001AuditingPracticesGroup

注 [1)] **ISO** のウェブサイト（http://www.iso.org）から入手可能.

ISO 9000

Fourth edition 2015-9-15

JIS Q 9000

2015-11-20

Quality management systems —Fundamentals and vocabulary

品質マネジメントシステム—基本及び用語（抜粋）

Introduction

This International Standard provides the fundamental concepts, principles and vocabulary for quality management systems (QMS) and provides the foundation for other QMS standards. This International Standard is intended to help the user to understand the fundamental concepts, principles and vocabulary of quality management, in order to be able to effectively and efficiently implement a QMS and realize value from other QMS standards.

This International Standard proposes a well-defined QMS, based on a framework that integrates established fundamental concepts, principles, processes and resources related to quality, in order to help organizations realize their objectives. It is applicable to all organizations, regardless of

序文

この規格は，2015年に第4版として発行されたISO 9000を基に，技術的内容及び構成を変更することなく作成した日本工業規格である．

なお，この規格で点線の下線を施してある参考事項は，対応国際規格にはない事項である．

この規格は，品質マネジメントシステム（QMS）の基本概念，原則及び用語を示しており，また，他のQMS規格の基礎となるものである．この規格は，より効果的かつ効率的にQMSを実施し，他のQMS規格の価値を実現するために，利用者が品質マネジメントの基本概念，原則及び用語を理解するのに役立つことを意図している．

この規格は，組織がその目標を実現するのを助けるために，確立された品質に関する基本概念，原則，プロセス及び資源を統合する枠組みに基づく，明確に定義されたQMSを示している．この規格は，組織の規模，複雑さ又はビジネスモデルを問わず，全ての組織に適用できる．その狙いは，製品及びサー

size, complexity or business model. Its aim is to increase an organization's awareness of its duties and commitment in fulfilling the needs and expectations of its customers and interested parties, and in achieving satisfaction with its products and services.

This International Standard contains seven quality management principles supporting the fundamental concepts described in 2.2. In 2.3, for each quality management principle, there is a "statement" describing each principle, a "rationale" explaining why the organization would address the principle, "key benefits" that are attributed to the principles, and "possible actions" that an organization can take in applying the principle.

This International Standard contains the terms and definitions that apply to all quality management and QMS standards developed by ISO/TC 176, and other sector-specific QMS standards based on those standards, at the time of publication. The terms and definitions are arranged in conceptual order, with an alphabetical index pro-

ビスによって，顧客及び利害関係者のニーズ及び期待を満たし，満足を達成するという責務及びコミットメントに対する組織の認識を高めることにある．

この規格は，**2.2** に規定した基本概念を支援する七つの品質マネジメントの原則を含んでいる．**2.3** には，それぞれの品質マネジメントの原則に対して，各原則の“説明”，組織がその原則に取り組む理由を説明する“根拠”，その原則からもたらされる“主な便益”，及びその原則の適用に際し組織が“取り得る行動”を規定している．

この規格は，この規格の発効時点における，**ISO/TC 176**（品質管理及び品質保証）によって作成された全ての品質マネジメント規格及びQMS規格，並びにそれらの規格に基づくセクター別QMS規格に適用される用語及び定義を含んでいる．用語及び定義は，概念の順に配列し，巻末には五十音順及びアルファベット順の索引を記載した．**附属書 A**

vided at the end of the document. Annex A includes a set of diagrams of the concept systems that form the concept ordering.

NOTE Guidance on some additional frequently-used words in the QMS standards developed by ISO/TC 176, and which have an identified dictionary meaning, is provided in a glossary available at: http://www.iso.org/iso/03_terminology_used_in_iso_9000_family.pdf

1 Scope

This International Standard describes the fundamental concepts and principles of quality management which are universally applicable to the following:

— organizations seeking sustained success through the implementation of a quality management system;

— customers seeking confidence in an organization's ability to consistently provide products and services conforming to their requirements;

— organizations seeking confidence in their

には，概念の順序を形成する一連の概念の体系図を示した．

（附属書 A は本書への収録を省略しています．）

注記　**ISO/TC 176** によって作成された QMS 規格に頻繁に用いられ，かつ，特定の辞書的な意味をもつ幾つかの追加的な言葉の手引は，次の URL に示された用語集にある．

http://www.iso.org/iso/03_terminology_used_in_iso_9000_family.pdf

1　適用範囲

この規格は，次の組織及び人に広く適用できる，品質マネジメントの基本概念及び原則について規定する．

— 品質マネジメントシステムの実施を通して持続的成功を求める組織

— 要求事項に適合する製品及びサービスを一貫して提供するための組織の能力について，信頼感を得ようとする顧客

— 製品及びサービスの要求事項が満たされるとい

supply chain that product and service requirements will be met;

— organizations and interested parties seeking to improve communication through a common understanding of the vocabulary used in quality management;
— organizations performing conformity assessments against the requirements of ISO 9001;
— providers of training, assessment or advice in quality management;
— developers of related standards.

This International Standard specifies the terms and definitions that apply to all quality management and quality management system standards developed by ISO/TC 176.

う信頼感を，自らのサプライチェーンにおいて得ようとする組織

— 品質マネジメントで用いる用語の共通理解を通して，コミュニケーションを改善しようとする組織及び利害関係者

— **JIS Q 9001** の要求事項に対する適合性評価を行う組織

— 品質マネジメントに関する教育・訓練，評価又は助言の提供者

— 関連する規格の作成者

また，この規格は，**ISO/TC 176** によって作成された全ての品質マネジメント及び品質マネジメントシステム規格に適用される用語及び定義も規定している．

注記　この規格の対応国際規格及びその対応の程度を表す記号を，次に示す．

ISO 9000:2015，Quality management systems — Fundamentals and vocabulary（IDT）

なお，対応の程度を表す記号“IDT”は，**ISO/IEC Guide 21-1** に基づき，“一致している”ことを示す．

2 Fundamental concepts and quality management principles

2.1 General

The quality management concepts and principles described in this International Standard give the organization the capacity to meet challenges presented by an environment that is profoundly different from recent decades. The context in which an organization works today is characterized by accelerated change, globalization of markets and the emergence of knowledge as a principal resource. The impact of quality extends beyond customer satisfaction: it can also have a direct impact on the organization's reputation.

Society has become better educated and more demanding, making interested parties increasingly more influential. By providing fundamental concepts and principles to be used in the development of a quality management system (QMS), this International Standard provides a way of thinking about the organization more broadly.

All concepts, principles and their interrelation-

2　基本概念及び品質マネジメントの原則

2.1　一般

この規格に規定する品質マネジメントの概念及び原則は，組織に，ここ数十年とは本質的に異なる環境からもたらされる課題に立ち向かう能力を与える．今日，組織が置かれている状況は，急速な変化，市場のグローバル化及び主要な資源としての知識の出現によって特徴付けられる．品質の影響は，顧客満足を超えた範囲にまでわたり，そうした影響が，組織の評判に直接影響を与えることもある．

社会においては，教育水準が上がり，要求が厳しくなり，利害関係者の影響力がますます強くなっている．この規格は，QMS の構築・発展に用いる基本概念及び原則を示すことによって，より広範に組織についての考え方を提供する．

全ての概念及び原則並びにそれらの相互関係は，

ships should be seen as a whole and not in isolation of each other. No individual concept or principle is more important than another. At any one time, finding the right balance in application is critical.

2.2 Fundamental concepts

2.2.1 Quality

An organization focused on quality promotes a culture that results in the behaviour, attitudes, activities and processes that deliver value through fulfilling the needs and expectations of customers and other relevant interested parties.

The quality of an organization's products and services is determined by the ability to satisfy customers and the intended and unintended impact on relevant interested parties.

The quality of products and services includes not only their intended function and performance, but also their perceived value and benefit to the customer.

全体として捉えるのがよく，それぞれを切り離して捉えないほうがよい．ある概念又は原則が，もう一つの概念又は原則よりも重要だということはない．いかなる場合にも，適用における適切なバランスを見つけることが重要である．

2.2　基本概念

2.2.1　品質

品質を重視する組織は，顧客及びその他の密接に関連する利害関係者のニーズ及び期待を満たすことを通じて価値を提供する行為，態度，活動及びプロセスをもたらすような文化を促進する．

ある組織の製品及びサービスの品質は，顧客を満足させる能力，並びに密接に関連する利害関係者に対する意図した影響及び意図しない影響によって決まる．

製品及びサービスの品質には，意図した機能及びパフォーマンスだけでなく，顧客によって認識された価値及び顧客に対する便益も含まれる．

2.2.2 Quality management system

A QMS comprises activities by which the organization identifies its objectives and determines the processes and resources required to achieve desired results.

The QMS manages the interacting processes and resources required to provide value and realize results for relevant interested parties.

The QMS enables top management to optimize the use of resources considering the long and short term consequences of their decision.

A QMS provides the means to identify actions to address intended and unintended consequences in providing products and services.

2.2.3 Context of an organization

Understanding the context of the organization is a process. This process determines factors which influence the organization's purpose, objectives and sustainability. It considers internal factors such as values, culture, knowledge and perfor-

2.2.2　品質マネジメントシステム

QMS は，組織が自らの目標を特定する活動，並びに組織が望む結果を達成するために必要なプロセス及び資源を定める活動から成る．

QMS は，密接に関連する利害関係者に価値を提供し，かつ，結果を実現するために必要な，相互に作用するプロセス及び資源をマネジメントする．

QMS によって，トップマネジメントは，自らの決定の長期的及び短期的な結果を考慮しながら，資源の利用を最適化することができる．

QMS は，製品及びサービスの提供において，意図した結果及び意図しない結果に取り組むための処置を特定する手段を提供する．

2.2.3　組織の状況

組織の状況を理解することは，一つのプロセスである．このプロセスにおいては，組織の目的，目標及び持続可能性に影響を与える要因を明確にする．また，組織の価値観，文化，知識，パフォーマンスなどの内部要因を考慮する．さらに，法的環境，技

mance of the organization. It also considers external factors such as legal, technological, competitive, market, cultural, social and economic environments.

Examples of the ways in which an organization's purpose can be expressed include its vision, mission, policies and objectives.

2.2.4 Interested parties

The concept of interested parties extends beyond a focus solely on the customer. It is important to consider all relevant interested parties.

Part of the process for understanding the context of the organization is to identify its interested parties. The relevant interested parties are those that provide significant risk to organizational sustainability if their needs and expectations are not met. Organizations define what results are necessary to deliver to those relevant interested parties to reduce that risk.

Organizations attract, capture and retain the

術的環境，競争環境，市場環境，文化的環境，社会的環境，経済的環境などの外部要因も考慮する．

組織の目的は，例えば，組織のビジョン，使命，方針及び目標を通じて表明することができる．

2.2.4　利害関係者

利害関係者の概念は，顧客だけを重要視するという考え方を超えるものである．密接に関連する利害関係者全てを考慮することが重要である．

組織の状況を理解するためのプロセスの一部として，その利害関係者を特定する．密接に関連する利害関係者とは，そのニーズ及び期待が満たされない場合に，組織の持続可能性に重大なリスクを与える利害関係者である．組織は，そうしたリスクを低減するために，これらの密接に関連する利害関係者に対して提供する必要がある結果は何かを定義する．

組織は，自らの成功を左右する密接に関連する利

support of the relevant interested parties they depend upon for their success.

2.2.5 Support

2.2.5.1 General

Top management support of the QMS and engagement of people enables:

- provision of adequate human and other resources;
- monitoring processes and results;
- determining and evaluating of risks and opportunities;
- implementing appropriate actions.

Responsible acquisition, deployment, maintenance, enhancement and disposal of resources support the organization in achieving its objectives.

2.2.5.2 People

People are essential resources within the organization. The performance of the organization is dependent upon how people behave within the system in which they work.

害関係者の支援を誘引し，獲得し，これを保持する．

2.2.5　支援

2.2.5.1　一般

QMSへのトップマネジメントの支援及び人々の積極的参加によって，次の事項が可能となる．

— 十分な人的資源及びその他の資源の提供

— プロセス及び結果の監視

— リスク及び機会の明確化及び評価

— 適切な処置の実施

責任をもって資源を取得し，展開し，維持し，増強し，処分・処遇することで，組織がその目標を達成することを支援する．

2.2.5.2　人々

人々は，組織内において欠かせない資源である．組織のパフォーマンスは，人々が，各自が働いているシステムの中でどのように行動するかによって決まる．

Within an organization, people become engaged and aligned through a common understanding of the quality policy and the organization's desired results.

2.2.5.3 Competence

A QMS is most effective when all employees understand and apply the skills, training, education and experience needed to perform their roles and responsibilities. It is the responsibility of top management to provide opportunities for people to develop these necessary competencies.

2.2.5.4 Awareness

Awareness is attained when people understand their responsibilities and how their actions contribute to the achievement of the organization's objectives.

2.2.5.5 Communication

Planned and effective internal (i.e. throughout the organization) and external (i.e. with relevant interested parties) communication enhances people's engagement and increased understanding

組織内において，人々は，品質方針及び組織が望む結果についての共通の理解を通して，積極的に参加し，連携するようになる．

2.2.5.3　力量

全ての従業員が，各自の役割及び責任を果たすために必要な技能，訓練，教育及び経験を理解し，これを適用したとき，QMSは最も効果的なものとなる．これらの必要な力量を身に付ける機会を人々に与えることは，トップマネジメントの責任である．

2.2.5.4　認識

人々が，各自の責任を理解し，自らの行動が組織の目標の達成にどのように貢献するかを理解したとき，認識は確固としたものになる．

2.2.5.5　コミュニケーション

計画された効果的な内部（すなわち，組織全体にわたる）コミュニケーション及び外部（すなわち，密接に関連する利害関係者との）コミュニケーションは，人々の積極的参加を増大させ，次の事項に対

of:

— the context of the organization;

— the needs and expectations of customers and other relevant interested parties;

— the QMS.

2.3 Quality management principles

2.3.1 Customer focus

2.3.1.1 Statement

The primary focus of quality management is to meet customer requirements and to strive to exceed customer expectations.

2.3.1.2 Rationale

Sustained success is achieved when an organization attracts and retains the confidence of customers and other relevant interested parties. Every aspect of customer interaction provides an opportunity to create more value for the customer. Understanding current and future needs of customers and other interested parties contributes to the sustained success of the organization.

する理解を深める.

— 組織の状況
— 顧客及びその他の密接に関連する利害関係者のニーズ及び期待
— QMS

2.3　品質マネジメントの原則

2.3.1　顧客重視

2.3.1.1　説明

品質マネジメントの主眼は，顧客の要求事項を満たすこと及び顧客の期待を超える努力をすることにある.

2.3.1.2　根拠

持続的成功は，組織が顧客及びその他の密接に関連する利害関係者を引き付け，その信頼を保持することによって達成できる．顧客との相互作用のあらゆる側面が，顧客のために更なる価値を創造する機会を与える．顧客及びその他の利害関係者の現在及び将来のニーズを理解することは，組織の持続的成功に寄与する.

2.3.1.3 Key benefits

Some potential key benefits are:

— increased customer value;
— increased customer satisfaction;
— improved customer loyalty;
— enhanced repeat business;
— enhanced reputation of the organization;
— expanded customer base;
— increased revenue and market share.

2.3.1.4 Possible actions

Possible actions include:

— recognize direct and indirect customers as those who receive value from the organization;
— understand customers' current and future needs and expectations;
— link the organization's objectives to customer needs and expectations;
— communicate customer needs and expectations throughout the organization;
— plan, design, develop, produce, deliver and support products and services to meet customer needs and expectations;

2.3.1.3　主な便益

あり得る主な便益を，次に示す．

— 顧客価値の増加
— 顧客満足の増加
— 顧客のロイヤリティの改善
— リピートビジネスの増加
— 組織の評判の向上
— 顧客基盤の拡大
— 収益及び市場シェアの増加

2.3.1.4　取り得る行動

取り得る行動を，次に示す．

— 直接的及び間接的な顧客を組織から価値を受け取る者として認識する．
— 顧客の現在及び将来のニーズ及び期待を理解する．
— 組織の目標を顧客のニーズ及び期待に関連付ける．
— 顧客のニーズ及び期待を組織全体に伝達する．
— 顧客のニーズ及び期待を満たす製品及びサービスを計画し，設計し，開発し，製造し，引き渡し，サポートする．

— measure and monitor customer satisfaction and take appropriate actions;
— determine and take action on relevant interested parties' needs and appropriate expectations that can affect customer satisfaction;
— actively manage relationships with customers to achieve sustained success.

2.3.2 Leadership

2.3.2.1 Statement

Leaders at all levels establish unity of purpose and direction and create conditions in which people are engaged in achieving the organization's quality objectives.

2.3.2.2 Rationale

Creation of unity of purpose and the direction and engagement of people enable an organization to align its strategies, policies, processes and resources to achieve its objectives.

2.3.2.3 Key benefits

Some potential key benefits are:

— increased effectiveness and efficiency in

— 顧客満足を測定・監視し，適切な処置をとる．

— 顧客満足に影響を与え得る密接に関連する利害関係者のニーズ及び適切な期待を明確にし，処置をとる．
— 持続的成功を達成するために，顧客との関係を積極的にマネジメントする．

2.3.2 リーダーシップ

2.3.2.1 説明

全ての階層のリーダーは，目的及び目指す方向を一致させ，人々が組織の品質目標の達成に積極的に参加している状況を作り出す．

2.3.2.2 根拠

目的及び目指す方向の一致並びに人々の積極的な参加によって，組織は，その目標の達成に向けて戦略，方針，プロセス及び資源を密接に関連付けることができる．

2.3.2.3 主な便益

あり得る主な便益を，次に示す．
— 組織の品質目標を満たす上での有効性及び効率

meeting the organization's quality objectives;
- better coordination of the organization's processes;
- improved communication between levels and functions of the organization;
- development and improvement of the capability of the organization and its people to deliver desired results.

2.3.2.4 Possible actions

Possible actions include:

- communicate the organization's mission, vision, strategy, policies and processes throughout the organization;
- create and sustain shared values, fairness and ethical models for behaviour at all levels of the organization;
- establish a culture of trust and integrity;
- encourage an organization-wide commitment to quality;
- ensure that leaders at all levels are positive examples to people in the organization;
- provide people with the required resources, training and authority to act with account-

の向上

— 組織内のプロセス間のより良い協調

— 組織内の階層間及び機能間のコミュニケーションの改善

— 望む結果を出せるような，組織及び人々の実現能力の開発及び向上

2.3.2.4　取り得る行動

取り得る行動を，次に示す．

— 組織の使命，ビジョン，戦略，方針及びプロセスを組織全体に周知する．

— 組織の全ての階層において，共通の価値基準，公正性及び倫理的模範を作り，持続させる．

— 信頼及び誠実さの文化を確立する．

— 品質に対する組織全体にわたるコミットメントを奨励する．

— 全ての階層のリーダーが，組織の人々にとって模範となることを確実にする．

— 人々に対し，説明責任（accountability）を意識して行動するために必要な，資源，教育・訓

ability;
— inspire, encourage and recognize the contribution of people.

2.3.3 Engagement of people

2.3.3.1 Statement

Competent, empowered and engaged people at all levels throughout the organization are essential to enhance the organization's capability to create and deliver value.

2.3.3.2 Rationale

In order to manage an organization effectively and efficiently, it is important to respect and involve all people at all levels. Recognition, empowerment and enhancement of competence facilitate the engagement of people in achieving the organization's quality objectives.

2.3.3.3 Key benefits

Some potential key benefits are:

— improved understanding of the organization's quality objectives by people in the organization and increased motivation to achieve

練及び権限を与える.

— 人々の貢献を鼓舞し，奨励し，認める.

2.3.3　人々の積極的参加

2.3.3.1　説明

組織内の全ての階層にいる，力量があり，権限を与えられ，積極的に参加する人々が，価値を創造し提供する組織の実現能力を強化するために必須である.

2.3.3.2　根拠

組織を効果的かつ効率的にマネジメントするためには，組織の全ての階層の全ての人々を尊重し，それらの人々の参加を促すことが重要である．貢献を認め，権限を与え，力量を向上させることによって，組織の品質目標達成への人々の積極的な参加が促進される.

2.3.3.3　主な便益

あり得る主な便益を，次に示す.

— 組織の品質目標に対する組織の人々の理解の向上，及びそれを達成するための意欲の向上

them;
- enhanced involvement of people in improvement activities;
- enhanced personal development, initiatives and creativity;
- enhanced people satisfaction;
- enhanced trust and collaboration throughout the organization;
- increased attention to shared values and culture throughout the organization.

2.3.3.4 Possible actions

Possible actions include:
- communicate with people to promote understanding of the importance of their individual contribution;
- promote collaboration throughout the organization;
- facilitate open discussion and sharing of knowledge and experience;
- empower people to determine constraints to performance and to take initiatives without fear;
- recognize and acknowledge people's contribu-

— 改善活動における人々の参画の増大

— 個人の成長，主導性及び創造性の強化

— 人々の満足の増大

— 組織全体における信頼及び協力の増大

— 組織全体における共通の価値基準及び文化に対する注目の高まり

2.3.3.4　取り得る行動

取り得る行動を，次に示す．

— 各人の貢献の重要性の理解を促進するために，人々とコミュニケーションを行う．

— 組織全体で協力を促進する．

— オープンな議論，並びに知識及び経験の共有を促す．

— 人々が，パフォーマンスに関わる制約条件を明確にし，恐れることなく率先して行動できるよう，権限を与える．

— 人々の貢献，学習及び向上を認め，褒める．

tion, learning and improvement;
- enable self-evaluation of performance against personal objectives;
- conduct surveys to assess people's satisfaction, communicate the results and take appropriate actions.

2.3.4 Process approach

2.3.4.1 Statement

Consistent and predictable results are achieved more effectively and efficiently when activities are understood and managed as interrelated processes that function as a coherent system.

2.3.4.2 Rationale

The QMS consists of interrelated processes. Understanding how results are produced by this system enables an organization to optimize the system and its performance.

2.3.4.3 Key benefits

Some potential key benefits are:
- enhanced ability to focus effort on key processes and opportunities for improvement;

— 個人の目標に対するパフォーマンスの自己評価を可能にする.
— 人々の満足を評価し，その結果を伝達し，適切な処置をとるための調査を行う.

2.3.4　プロセスアプローチ

2.3.4.1　説明

活動を，首尾一貫したシステムとして機能する相互に関連するプロセスであると理解し，マネジメントすることによって，矛盾のない予測可能な結果が，より効果的かつ効率的に達成できる.

2.3.4.2　根拠

QMS は，相互に関連するプロセスで構成される.このシステムによって結果がどのように生み出されるかを理解することで，組織は，システム及びそのパフォーマンスを最適化できる.

2.3.4.3　主な便益

あり得る主な便益を，次に示す.
— 主要なプロセス及び改善のための機会に注力する能力の向上

— consistent and predictable outcomes through a system of aligned processes;

— optimized performance through effective process management, efficient use of resources and reduced cross-functional barriers;
— enabling the organization to provide confidence to interested parties related to its consistency, effectiveness and efficiency.

2.3.4.4 Possible actions

Possible actions include:
— define objectives of the system and processes necessary to achieve them;
— establish authority, responsibility and accountability for managing processes;
— understand the organization's capabilities and determine resource constraints prior to action;
— determine process interdependencies and analyse the effect of modifications to individual processes on the system as a whole;
— manage processes and their interrelations as a system to achieve the organization's quali-

— 密接に関連付けられたプロセスから構成されるシステムを通して得られる矛盾のない，予測可能な成果
— 効果的なプロセスのマネジメント，資源の効率的な利用，及び機能間の障壁の低減を通して得られるパフォーマンスの最適化
— 組織に整合性があり，有効でかつ効率的であることに関して利害関係者に信頼感を与えることができるようになる.

2.3.4.4　取り得る行動

取り得る行動を，次に示す.

— システムの目標，及びそれらを達成するために必要なプロセスを定める.
— プロセスをマネジメントするための権限，責任及び説明責任（accountability）を確立する.
— 組織の実現能力を理解し，実行前に資源の制約を明確にする.

— プロセスの相互依存関係を明確にし，システム全体で個々のプロセスへの変更の影響を分析する.
— 組織の品質目標を効果的及び効率的に達成するために，プロセス及びその相互関係をシステム

ty objectives effectively and efficiently;

— ensure the necessary information is available to operate and improve the processes and to monitor, analyse and evaluate the performance of the overall system;

— manage risks which can affect outputs of the processes and overall outcomes of the QMS.

2.3.5 Improvement

2.3.5.1 Statement

Successful organizations have an ongoing focus on improvement.

2.3.5.2 Rationale

Improvement is essential for an organization to maintain current levels of performance, to react to changes in its internal and external conditions and to create new opportunities.

2.3.5.3 Key benefits

Some potential key benefits are:

— improved process performance, organizational capability and customer satisfaction;

— enhanced focus on root cause investigation

としてマネジメントする．

— プロセスを運用し，改善するとともに，システム全体のパフォーマンスを監視し，分析し，評価するために必要な情報が利用できる状態にあることを確実にする．
— プロセスのアウトプット及び QMS の全体的な成果に影響を与え得るリスクを管理する．

2.3.5　改善

2.3.5.1　説明

成功する組織は，改善に対して，継続して焦点を当てている．

2.3.5.2　根拠

改善は，組織が，現レベルのパフォーマンスを維持し，内外の状況の変化に対応し，新たな機会を創造するために必須である．

2.3.5.3　主な便益

あり得る主な便益を，次に示す．

— プロセスパフォーマンス，組織の実現能力及び顧客満足の改善
— 予防及び是正処置につながる根本原因の調査及

and determination, followed by prevention and corrective actions;

- enhanced ability to anticipate and react to internal and external risks and opportunities;
- enhanced consideration of both incremental and breakthrough improvement;
- improved use of learning for improvement;
- enhanced drive for innovation.

2.3.5.4 Possible actions

Possible actions include:

- promote establishment of improvement objectives at all levels of the organization;
- educate and train people at all levels on how to apply basic tools and methodologies to achieve improvement objectives;
- ensure people are competent to successfully promote and complete improvement projects;
- develop and deploy processes to implement improvement projects throughout the organization;
- track, review and audit the planning, implementation, completion and results of im-

び確定の重視

— 内部及び外部のリスク及び機会を予測し，これに対応するための能力の強化

— 漸増的な改善と飛躍的な改善の両方に関する検討の強化
— 改善のための学習に関する工夫
— 革新に対する意欲の向上

2.3.5.4　取り得る行動

取り得る行動を，次に示す．
— 組織の全ての階層において改善目標の設定を促す．
— 改善目標を達成するための基本的なツール及び方法論の適用の仕方に関し，全ての階層の人々に教育及び訓練を行う．
— 改善プロジェクトを成功裏に促進し，完結するための力量を人々がもつことを確実にする．
— 組織全体で改善プロジェクトを実施するためのプロセスを開発し，展開する．

— 改善プロジェクトの計画，実施，完了及び結果を追跡し，レビューし，監査する．

provement projects;

— integrate improvement consideration into development of new or modified products and services and processes;

— recognize and acknowledge improvement.

2.3.6 Evidence-based decision making

2.3.6.1 Statement

Decisions based on the analysis and evaluation of data and information are more likely to produce desired results.

2.3.6.2 Rationale

Decision-making can be a complex process and it always involves some uncertainty. It often involves multiple types and sources of inputs, as well as their interpretation, which can be subjective. It is important to understand cause and effect relationships and potential unintended consequences. Facts, evidence and data analysis lead to greater objectivity and confidence in decision making.

2.3.6.3 Key benefits

Some potential key benefits are:

— 新規の又は変更された，製品及びサービス並びにプロセスの開発に，改善の考えを組み込む．

— 改善を認め，褒める．

2.3.6　客観的事実に基づく意思決定

2.3.6.1　説明

データ及び情報の分析及び評価に基づく意思決定によって，望む結果が得られる可能性が高まる．

2.3.6.2　根拠

意思決定は，複雑なプロセスとなる可能性があり，常に何らかの不確かさを伴う．意思決定は，主観的かもしれない，複数の種類の，複数の源泉からのインプット，及びそれらに対する解釈を含むことが多い．因果関係，及び起こり得る意図しない帰結を理解することが重要である．客観的事実，根拠及びデータ分析は，意思決定の客観性及び信頼性を高める．

2.3.6.3　主な便益

あり得る主な便益を，次に示す．

— improved decision making processes;
— improved assessment of process performance and ability to achieve objectives;
— improved operational effectiveness and efficiency;
— increased ability to review, challenge and change opinions and decisions;
— increased ability to demonstrate the effectiveness of past decisions.

2.3.6.4 Possible actions

Possible actions include:

— determine, measure and monitor key indicators to demonstrate the organization's performance;
— make all data needed available to the relevant people;
— ensure that data and information are sufficiently accurate, reliable and secure;
— analyse and evaluate data and information using suitable methods;
— ensure people are competent to analyse and evaluate data as needed;
— make decisions and take actions based on ev-

— 意思決定プロセスの改善
— プロセスパフォーマンスの評価及び目標の達成能力の改善
— 運用の有効性及び効率の改善

— 意見及び決定をレビューし，異議を唱え，変更する能力の向上
— 過去の決定の有効性を実証する能力の向上

2.3.6.4 取り得る行動

取り得る行動を，次に示す．

— 組織のパフォーマンスを示す主な指標を決定し，測定し，監視する．

— 全ての必要なデータを，関連する人々が利用できる状態にする．
— データ及び情報が十分に正確で，信頼性があり，安全であることを確実にする．
— データ及び情報を，適切な方法を用いて分析し，評価する．
— 人々が，必要に応じてデータを分析し，評価する力量をもつことを確実にする．
— 経験と勘とのバランスがとれた意思決定を行

idence, balanced with experience and intuition.

2.3.7 Relationship management

2.3.7.1 Statement

For sustained success, organizations manage their relationships with relevant interested parties, such as providers.

2.3.7.2 Rationale

Relevant interested parties influence the performance of an organization. Sustained success is more likely to be achieved when the organization manages relationships with all of its interested parties to optimize their impact on its performance. Relationship management with its provider and partner networks is of particular importance.

2.3.7.3 Key benefits

Some potential key benefits are:

— enhanced performance of the organization and its relevant interested parties through responding to the opportunities and con-

い，客観的事実に基づいた処置をとる．

2.3.7　関係性管理

2.3.7.1　説明

持続的成功のために，組織は，例えば提供者のような，密接に関連する利害関係者との関係をマネジメントする．

2.3.7.2　根拠

密接に関連する利害関係者は，組織のパフォーマンスに影響を与える．持続的成功は，組織のパフォーマンスに対する利害関係者の影響を最適化するように全ての利害関係者との関係をマネジメントすると，より達成しやすくなる．提供者及びパートナとのネットワークにおける関係性管理は特に重要である．

2.3.7.3　主な便益

あり得る主な便益を，次に示す．

— それぞれの利害関係者に関連する機会及び制約に対応することを通じた，組織及びその密接に関連する利害関係者のパフォーマンスの向上

straints related to each interested party;
— common understanding of objectives and values among interested parties;
— increased capability to create value for interested parties by sharing resources and competence and managing quality related risks;
— a well-managed supply chain that provides a stable flow of products and services.

2.3.7.4 Possible actions

Possible actions include:
— determine relevant interested parties (such as providers, partners, customers, investors, employees or society as a whole) and their relationship with the organization;
— determine and prioritize interested party relationships that need to be managed;
— establish relationships that balance short-term gains with long-term considerations;
— gather and share information, expertise and resources with relevant interested parties;
— measure performance and provide performance feedback to interested parties, as appropriate, to enhance improvement initia-

— 利害関係者の目標及び価値観に関する共通理解

— 資源及び力量の共有，並びに品質関連のリスクの管理による，利害関係者のための価値を創造する実現能力の向上
— 製品及びサービスの安定した流れを提供する，よく管理されたサプライチェーン

2.3.7.4 取り得る行動

取り得る行動を，次に示す．
— 密接に関連する利害関係者（例えば，提供者，パートナ，顧客，投資者，従業員，社会全体）及びそれらの組織との関係を明確にする．

— マネジメントする必要のある利害関係者との関係性を明確にし，優先順位を付ける．
— 短期的な利益と長期的な考慮とのバランスがとれた関係を構築する．
— 情報，専門的知識及び資源を，密接に関連する利害関係者との間で収集し，共有する．
— 改善の取組みを強化するために，適切に，パフォーマンスを測定し，利害関係者に対してフィードバックを行う．

tives;

— establish collaborative development and improvement activities with providers, partners and other interested parties;

— encourage and recognize improvements and achievements by providers and partners.

2.4 Developing the QMS using fundamental concepts and principles

2.4.1 QMS model

2.4.1.1 General

Organizations share many characteristics with humans as a living and learning social organism. Both are adaptive and comprise interacting systems, processes and activities. In order to adapt to their varying context, each needs the ability to change. Organizations often innovate to achieve breakthrough improvements. An organization's QMS model recognizes that not all systems, processes and activities can be predetermined; therefore it needs to be flexible and adaptable within the complexities of the organizational context.

— 提供者，パートナ及びその他の利害関係者と協力して開発及び改善活動を行う．

— 提供者及びパートナによる改善及び達成を奨励し，認める．

2.4 基本概念及び原則を用いたQMSの構築・発展

2.4.1 QMSモデル

2.4.1.1 一般

組織は，学習する生きた社会有機体としての人間と多くの特性を共有している．両者とも適応性をもち，相互に作用するシステム，プロセス及び活動で構成されている．また，両者とも，様々な状況に適応するために，変化する能力が必要である．組織は，飛躍的な改善を達成するために，革新を行うことが多い．組織のQMSモデルについては，全てのシステム，プロセス及び活動をあらかじめ決定できるとは限らないことが分かっている．したがって，組織のQMSモデルは，組織の状況の複雑さの中で，柔軟性及び順応性を備えたものであることが必要である．

2.4.1.2 System

Organizations seek to understand the internal and external context to identify the needs and expectations of relevant interested parties. This information is used in the development of the QMS to achieve organizational sustainability. The outputs from one process can be the inputs into other processes and are interlinked into the overall network. Although often appearing to be comprised of similar processes, each organization and its QMS is unique.

2.4.1.3 Processes

The organization has processes that can be defined, measured and improved. These processes interact to deliver results consistent with the organization's objectives and cross functional boundaries. Some processes can be critical while others are not. Processes have interrelated activities with inputs to deliver outputs.

2.4.1.4 Activity

People collaborate within a process to carry out

2.4.1.2　システム

組織は，密接に関連する利害関係者のニーズ及び期待を特定するために，内部及び外部の状況を理解しようとする．この情報は，組織の持続可能性を達成するための QMS を構築・発展する上で用いられる．あるプロセスからのアウトプットは，他のプロセスへのインプットとなり得るものであり，相互に結び付いて全体的なネットワークを形成する．類似のプロセスで構成されているように見えることが多いが，各組織及びその QMS は固有のものである．

2.4.1.3　プロセス

組織には，定義，測定及び改善することができるプロセスが存在する．これらのプロセスは，組織の目標と整合した結果をもたらすように，また，機能間の壁を越えるように相互に作用する．プロセスには，それほど重要でないものもあるが，重要なものもある．プロセスでは，インプットを用いてアウトプットを出すための相互に関連する活動が行われる．

2.4.1.4　活動

人々は，プロセスの中で業務を行うために協力し，

their daily activities. Some activities are prescribed and depend on an understanding of the objectives of the organization, while others are not and react to external stimuli to determine their nature and execution.

2.4.2 Development of a QMS

A QMS is a dynamic system that evolves over time through periods of improvement. Every organization has quality management activities, whether they have been formally planned or not. This International Standard provides guidance on how to develop a formal system to manage these activities. It is necessary to determine activities which already exist in the organization and their suitability regarding the context of the organization. This International Standard, along with ISO 9004 and ISO 9001, can then be used to assist the organization to develop a cohesive QMS.

A formal QMS provides a framework for planning, executing, monitoring and improving the performance of quality management activities. The QMS does not need to be complicated; rather

各自の日々の活動を行う．活動の中には，あらかじめ定められ，組織の目標についての理解によって決まるものがあるが，他方で，あらかじめ定められておらず，外部からの刺激に応じてその性質及び実行を決定するものもある．

2.4.2　QMS の構築・発展

QMS は，複数の改善の時期を経て，時とともに進化する，動的なシステムである．各組織は，正式に計画したものかどうかにかかわらず，品質マネジメント活動を行っている．この規格は，これらの活動をマネジメントするための正式なシステムをどのように開発するかについての手引を提供する．まず，既に組織内に存在する活動，及び組織の状況に関するその適切性を明確にする必要がある．この規格，**JIS Q 9004** 及び **JIS Q 9001** は，そうした後で，組織がまとまりのある QMS を構築・発展するのを助けるために用いることができる．

正規の QMS は，品質マネジメント活動のパフォーマンスを計画し，実行し，監視し，改善するための枠組みを提供する．QMS は，複雑である必要はなく，むしろ組織のニーズを正確に反映してい

it needs to accurately reflect the needs of the organization. In developing the QMS, the fundamental concepts and principles given in this International Standard can provide valuable guidance.

QMS planning is not a singular event, rather it is an ongoing process. Plans evolve as the organization learns and circumstances change. A plan takes into account all quality activities of the organization and ensures that it covers all guidance of this International Standard and requirements of ISO 9001. The plan is implemented upon approval.

It is important for an organization to regularly monitor and evaluate both the implementation of the plan and the performance of the QMS. Carefully considered indicators facilitate these monitoring and evaluation activities.

Auditing is a means of evaluating the effectiveness of the QMS, in order to identify risks and to determine the fulfilment of requirements. In or-

る必要がある．QMS を構築・発展するに当たって，この規格に示す基本概念及び原則は，貴重な手引となり得る．

QMS の計画策定は，一回限りの事象ではなく，むしろ継続するプロセスである．計画は，組織が学習し，周辺状況が変化するにつれて進化する．計画は，組織の全ての品質活動を考慮しており，また，この規格の全ての指針及び **JIS Q 9001** の要求事項を網羅することを確実にすることが望ましい．計画は，承認を受けて実施する．

組織にとって重要なことは，計画の実施と QMS のパフォーマンスの両方を，定期的に監視及び評価することである．入念に検討された指標によって，この監視及び評価の活動が容易となる．

監査は，リスクを特定し，要求事項を満たしていることを明確にするために，QMS の有効性を評価する手段である．監査を有効なものにするために，

der for audits to be effective, tangible and intangible evidence needs to be collected. Actions are taken for correction and improvement based upon analysis of the evidence gathered. The knowledge gained could lead to innovation, taking QMS performance to higher levels.

2.4.3 QMS standards, other management systems and excellence models

The approaches to a QMS described in QMS standards developed by ISO/TC 176, in other management system standards and in organizational excellence models are based on common principles. They all enable an organization to identify risks and opportunities and contain guidance for improvement. In the current context, many issues such as innovation, ethics, trust and reputation could be regarded as parameters within the QMS. Standards related to quality management (e.g. ISO 9001), environmental management (e.g. ISO 14001) and energy management (e.g. ISO 50001), as well as other management standards and organizational excellence models, have addressed this.

有形及び無形の客観的事実を収集する必要がある．収集した客観的事実の分析に基づき，修正及び改善のために処置をとる．取得した知識が，QMSのパフォーマンスをより高いレベルに押し上げ，革新につながることがある．

2.4.3　QMSの規格，他のマネジメントシステム及び卓越モデル

ISO/TC 176によって作成されたQMS規格，他のマネジメントシステム規格及び組織の卓越モデルに示されているQMSのアプローチは，共通の原則に基づいている．これら全ては，組織がリスク及び機会を特定することを可能にし，改善のための手引を含んでいる．今日の状況では，革新，倫理，信頼及び評判のような多くの課題は，QMSにおいて考慮すべき属性とみなされ得る．品質マネジメントに関する規格（**ISO 9001**など），環境マネジメントに関する規格（**ISO 14001**など），エネルギーマネジメントに関する規格（**ISO 50001**など）及びその他のマネジメント規格，並びに組織の卓越モデルがこれに対応してきた．

The QMS standards developed by ISO/TC 176 provide a comprehensive set of requirements and guidelines for a QMS. ISO 9001 specifies requirements for a QMS. ISO 9004 provides guidance on a wide range of objectives of a QMS for sustainable success and improved performance. Guidelines for components of a QMS include ISO 10001, ISO 10002, ISO 10003, ISO 10004, ISO 10008, ISO 10012 and ISO 19011. Guidelines for technical subjects in support of a QMS include ISO 10005, ISO 10006, ISO 10007, ISO 10014, ISO 10015, ISO 10018 and ISO 10019. Technical reports in support of a QMS include ISO/TR 10013 and ISO/TR 10017. Requirements for a QMS are also provided in sector-specific standards, such as ISO/TS 16949.

The various parts of an organization's management system, including its QMS, can be integrated as a single management system. The objectives, processes and resources related to quality, growth, funding, profitability, environment, occupational health and safety, energy, security and other aspects of the organization can be more ef-

ISO/TC 176 によって作成された QMS 規格は，QMS に関する一連の包括的な要求事項及び指針を示す．**ISO 9001** は，QMS に関する要求事項を規定している．**ISO 9004** は，持続可能な成功及びパフォーマンス改善のための広範な QMS の目標に関する指針を示す．QMS の構成要素に関する指針には，**ISO 10001**～**ISO 10004**，**ISO 10008**，**ISO 10012** 及び **ISO 19011** がある．QMS を支援する技術的テーマに関する指針には，**ISO 10005**～**ISO 10007**，**ISO 10014**，**ISO 10015**，**ISO 10018** 及び **ISO 10019** がある．QMS を支援する技術報告書には，**ISO/TR 10013** 及び **ISO/TR 10017** がある．QMS に関する要求事項は，**ISO/TS 16949** などのセクター規格でも示されている．

組織のマネジメントシステムの様々な部分は，その QMS を含め，単一のマネジメントシステムとして統合することができる．QMS が他のマネジメントシステムと統合されたとき，品質，成長，資金調達，収益性，環境，労働安全衛生，エネルギー，セキュリティ及び組織のその他の側面に関する目標，プロセス及び資源を，より効果的かつ効率的に達成

fectively and efficiently achieved and used when the QMS is integrated with other management systems. The organization can perform an integrated audit of its management system against the requirements of multiple International Standards, such as ISO 9001, ISO 14001, ISO/IEC 27001 and ISO 50001.

NOTE The ISO handbook "The integrated use of management system standards" can provide useful guidance.

3 Terms and definitions

3.1 Terms related to person or people

3.1.1

top management

person or group of people who directs and controls an *organization* (3.2.1) at the highest level

Note 1 to entry: Top management has the power to delegate authority and provide resources within the organization.

Note 2 to entry: If the scope of the *management system* (3.5.3) covers only part of an organization,

し，利用することができる．組織は，**ISO 9001**，**ISO 14001**，**ISO/IEC 27001**，**ISO 50001** などの複数の規格の要求事項に照らして，自らのマネジメントシステムの統合監査を実施することができる．

注記　ISO ハンドブック“マネジメントシステム規格の統合的な利用”は，有益な手引を提供し得る．

3　用語及び定義

3.1　個人又は人々に関する用語

3.1.1

トップマネジメント（top management）

最高位で**組織**（**3.2.1**）を指揮し，管理する個人又はグループ．

注記 1　トップマネジメントは，組織内で，権限を委譲し，資源を提供する力をもっている．

注記 2　**マネジメントシステム**（**3.5.3**）の適用範囲が組織の一部だけの場合，トッ

then top management refers to those who direct and control that part of the organization.

Note 3 to entry: This constitutes one of the common terms and core definitions for ISO management system standards given in Annex SL of the Consolidated ISO Supplement to the ISO/IEC Directives, Part 1.

3.1.2
quality management system consultant

person who assists the *organization* (3.2.1) on *quality management system realization* (3.4.3), giving advice or *information* (3.8.2)

Note 1 to entry: The quality management system consultant can also assist in realizing parts of a *quality management system* (3.5.4).

Note 2 to entry: ISO 10019:2005 provides guidance on how to distinguish a competent quality management system consultant from one who is not competent.

プマネジメントとは，組織内のその一部を指揮し，管理する人をいう．

注記 3　この用語及び定義は，ISO/IEC 専門業務用指針―第 1 部：統合版 ISO 補足指針の**附属書 SL** に示された ISO マネジメントシステム規格の共通用語及び中核となる定義の一つを成す．

3.1.2

品質マネジメントシステムコンサルタント（quality management system consultant）

品質マネジメントシステムの実現（**3.4.3**）に関して，助言又は**情報**（**3.8.2**）を与えて**組織**（**3.2.1**）を支援する人．

注記 1　品質マネジメントシステムコンサルタントは，**品質マネジメントシステム**（**3.5.4**）の実現化の一部を支援することもできる．

注記 2　**JIS Q 10019**:2005 は，力量のある品質マネジメントシステムコンサルタントを，力量のないコンサルタントから区別するための手引を提供する．

[SOURCE: ISO 10019:2005, 3.2, modified]

3.1.3

involvement

taking part in an activity, event or situation

3.1.4

engagement

involvement (3.1.3) in, and contribution to, activities to achieve shared *objectives* (3.7.1)

3.1.5

configuration authority

configuration control board

dispositioning authority

person or a group of persons with assigned responsibility and authority to make decisions on the *configuration* (3.10.6)

Note 1 to entry: Relevant *interested parties* (3.2.3) within and outside the *organization* (3.2.1) should be represented on the configuration authority.

（**JIS Q 10019**:2005 の **3.2** を変更.）

3.1.3

参画（involvement）

活動，行事又は状況の一部を担うこと.

3.1.4

積極的参加（engagement）

共通の**目標**（**3.7.1**）を達成するために活動に**参画**（**3.1.3**）し，寄与すること.

3.1.5

コンフィギュレーション機関（configuration authority），**コンフィギュレーション統制委員会**（configuration control board），**コンフィギュレーション決定委員会**（dispositioning authority）

コンフィギュレーション（**3.10.6**）に関して，意思決定を行う責任及び権限を割り当てられた個人又はグループ.

注記　**組織**（**3.2.1**）内外の密接に関連する**利害関係者**（**3.2.3**）は，コンフィギュレーション機関に代表を派遣することが望ましい.

[SOURCE: ISO 10007:2003, 3.8, modified]

3.1.6

dispute resolver

<customer satisfaction> individual person assigned by a *DRP-provider* (3.2.7) to assist the parties in resolving a *dispute* (3.9.6)

EXAMPLE Staff, volunteer, *contract* (3.4.7) personnel.

[SOURCE: ISO 10003:2007, 3.7, modified]

3.2 Terms related to organization

3.2.1

organization

person or group of people that has its own functions with responsibilities, authorities and relationships to achieve its *objectives* (3.7.1)

Note 1 to entry: The concept of organization includes, but is not limited to, sole-trader, company, corporation, firm, enterprise, authority, partnership, *association* (3.2.8), charity or institution, or

（**ISO 10007**:2003 の **3.8** を変更.）

3.1.6

紛争解決者（dispute resolver）

＜顧客満足＞**紛争**（**3.9.6**）の解決において，当事者を援助するため，**DRP 提供者**（**3.2.7**）によって選定される個人.

例　スタッフ，ボランティア，**契約**（**3.4.7**）を結んだ個人

（**JIS Q 10003**:2010 の **3.7** を変更.）

3.2　組織に関する用語

3.2.1

組織（organization）

自らの**目標**（**3.7.1**）を達成するため，責任，権限及び相互関係を伴う独自の機能をもつ，個人又はグループ.

注記 1　組織という概念には，法人か否か，公的か私的かを問わず，自営業者，会社，法人，事務所，企業，当局，共同経営会社，**協会**（**3.2.8**），非営利団体若し

part or combination thereof, whether incorporated or not, public or private.

Note 2 to entry: This constitutes one of the common terms and core definitions for ISO management system standards given in Annex SL of the Consolidated ISO Supplement to the ISO/IEC Directives, Part 1. The original definition has been modified by modifying Note 1 to entry.

3.2.2

context of the organization

combination of internal and external issues that can have an effect on an *organization's* (3.2.1) approach to developing and achieving its *objectives* (3.7.1)

Note 1 to entry: The organization's objectives can be related to its *products* (3.7.6) and *services* (3.7.7), investments and behaviour towards its *interested parties* (3.2.3).

Note 2 to entry: The concept of context of the organization is equally applicable to not-for-profit or public service organizations as it is to those

くは機構，又はこれらの一部若しくは組合せが含まれる．ただし，これらに限定されるものではない．

注記 2　この用語及び定義は，ISO/IEC 専門業務用指針―第 1 部：統合版 ISO 補足指針の**附属書 SL** に示された ISO マネジメントシステム規格の共通用語及び中核となる定義の一つを成す．元の定義の**注記 1** を変更した．

3.2.2

組織の状況（context of the organization）

組織（**3.2.1**）がその**目標**（**3.7.1**）設定及び達成に向けて取るアプローチに影響を及ぼし得る，内部及び外部の課題の組合せ．

注記 1　組織の目標は，その**製品**（**3.7.6**）及び**サービス**（**3.7.7**），投資，並びに**利害関係者**（**3.2.3**）に対する行動に関連し得る．

注記 2　組織の状況という概念は，非営利又は公共サービスの組織に対しても，営利組織に対する場合と同様に適用でき

seeking profits.

Note 3 to entry: In English, this concept is often referred to by other terms such as "business environment", "organizational environment" or "ecosystem of an organization".

Note 4 to entry: Understanding the *infrastructure* (3.5.2) can help to define the context of the organization.

3.2.3

interested party

stakeholder

person or *organization* (3.2.1) that can affect, be affected by, or perceive itself to be affected by a decision or activity

EXAMPLE *Customers* (3.2.4), owners, people in an organization, *providers* (3.2.5), bankers, regulators, unions, partners or society that can include competitors or opposing pressure groups.

Note 1 to entry: This constitutes one of the common terms and core definitions for ISO manage-

る．

注記 3　組織の状況という概念は，組織の“事業環境（business environment）”，“組織環境（organizational environment）”，“組織のエコシステム（eco-system）”などと言われる場合もある．

注記 4　**インフラストラクチャ**（**3.5.2**）を理解することは，組織の状況を定める上で役立ち得る．

3.2.3

利害関係者（interested party），**ステークホルダー**（stakeholder）

ある決定事項若しくは活動に影響を与え得るか，その影響を受け得るか，又はその影響を受けると認識している，個人又は**組織**（**3.2.1**）．

例　**顧客**（**3.2.4**），所有者，組織内の人々，**提供者**（**3.2.5**），銀行家，規制当局，組合，パートナ，社会（競争相手又は対立する圧力団体を含むこともある．）

注記　この用語及び定義は，ISO/IEC 専門業務用指針―第 1 部：統合版 ISO 補足指

ment system standards given in Annex SL of the Consolidated ISO Supplement to the ISO/IEC Directives, Part 1. The original definition has been modified by adding the Example.

3.2.4

customer

person or *organization* (3.2.1) that could or does receive a *product* (3.7.6) or a *service* (3.7.7) that is intended for or required by this person or organization

EXAMPLE Consumer, client, end-user, retailer, receiver of product or service from an internal *process* (3.4.1), beneficiary and purchaser.

Note 1 to entry: A customer can be internal or external to the organization.

3.2.5

provider

supplier

organization (3.2.1) that provides a *product* (3.7.6) or a *service* (3.7.7)

針の**附属書 SL** に示された ISO マネジメントシステム規格の共通用語及び中核となる定義の一つを成す．元の定義にない例を追加した．

3.2.4

顧客（customer）

個人若しくは**組織**（**3.2.1**）向け又は個人若しくは組織から要求される**製品**（**3.7.6**）・**サービス**（**3.7.7**）を，受け取る又はその可能性のある個人又は組織．

例　消費者，依頼人，エンドユーザ，小売業者，内部**プロセス**（**3.4.1**）からの製品又はサービスを受け取る人，受益者，購入者

注記　顧客は，組織の内部又は外部のいずれでもあり得る．

3.2.5

提供者（provider），**供給者**（supplier）

製品（**3.7.6**）又は**サービス**（**3.7.7**）を提供する**組織**（**3.2.1**）．

EXAMPLE Producer, distributor, retailer or vendor of a product or a service.

Note 1 to entry: A provider can be internal or external to the organization.
Note 2 to entry: In a contractual situation, a provider is sometimes called “contractor”.

3.2.6
external provider
external supplier
provider (3.2.5) that is not part of the *organization* (3.2.1)

EXAMPLE Producer, distributor, retailer or vendor of a *product* (3.7.6) or a *service* (3.7.7)

3.2.7
DRP-provider
dispute resolution process provider
person or *organization* (3.2.1) that supplies and operates an external *dispute* (3.9.6) resolution *process* (3.4.1)

例　製品又はサービスの生産者，流通者，小売業者又は販売者

注記 1　提供者は，組織の内部又は外部のいずれでもあり得る．

注記 2　契約関係においては，提供者は“契約者”と呼ばれる．

3.2.6

外部提供者（external provider），**外部供給者**（external supplier）

組織（**3.2.1**）の一部ではない**提供者**（**3.2.5**）．

例　**製品**（**3.7.6**）又は**サービス**（**3.7.7**）の生産者，流通者，小売業者又は販売者

3.2.7

DRP 提供者（DRP-provider），**紛争解決手続提供者**（dispute resolution process provider）

組織の外部における**紛争**（**3.9.6**）解決の**プロセス**（**3.4.1**）を提供し，運用する個人又は**組織**（**3.2.1**）．

Note 1 to entry: Generally, a DRP-provider is a legal entity, separate from the organization or person as an individual and the complainant. In this way, the attributes of independence and fairness are emphasized. In some situations, a separate unit is established within the organization to handle unresolved *complaints* (3.9.3).

Note 2 to entry: The DRP-provider *contracts* (3.4.7) with the parties to provide dispute resolution, and is accountable for *performance* (3.7.8). The DRP-provider supplies *dispute resolvers* (3.1.6). The DRP-provider also utilizes support, executive and other managerial staff to supply financial resources, clerical support, scheduling assistance, training, meeting rooms, supervision and similar functions.

Note 3 to entry: DRP-providers can take many forms including not-for-profit, for-profit and public entities. An *association* (3.2.8) can also be a DRP-provider.

Note 4 to entry: In ISO 10003:2007 instead of the term DRP-provider, the term "provider" is used.

注記 1 一般に，DRP 提供者は，組織又は個人，及び苦情申出者とは異なる法的主体である．このことによって，独立性及び公正性という性質が強化される．状況によっては，未解決の**苦情**（**3.9.3**）に対応するため，組織内に別の部署が設けられる場合がある．

注記 2 DRP 提供者は，紛争解決手続を提供するために両当事者と**契約**（**3.4.7**）を結び，その**パフォーマンス**（**3.7.8**）に対する責任を負う．DRP 提供者は，**紛争解決者**（**3.1.6**）を選定する．DRP 提供者はまた，支援要員，経営幹部及びその他の管理層を活用し，財源，事務的サポート，スケジュール調整の援助，教育・訓練，会議室，監督及びこれに類する職務を提供する．

注記 3 DRP 提供者は，非営利団体，営利団体及び公共団体を含め，様々な形態をとることができる．**協会**（**3.2.8**）もまた，DRP 提供者になり得る．

注記 4 **JIS Q 10003**:2010 では，“DRP 提供者”の代わりに，“提供者”という用語が用いられている．

[SOURCE: ISO 10003:2007, 3.9, modified]

3.2.8

association

<customer satisfaction> *organization* (3.2.1) consisting of member organizations or persons

[SOURCE: ISO 10003:2007, 3.1]

3.2.9

metrological function

functional unit with administrative and technical responsibility for defining and implementing the *measurement management system* (3.5.7)

[SOURCE: ISO 10012:2003, 3.6, modified]

3.3 Terms related to activity

3.3.1

improvement

activity to enhance *performance* (3.7.8)

Note 1 to entry: The activity can be recurring or singular.

（**JIS Q 10003**:2010 の **3.9** を変更.）

3.2.8

協会（association）

＜顧客満足＞組織又は個人を会員として構成された**組織**（**3.2.1**）.

（**JIS Q 10003**:2010 の **3.1** 参照）

3.2.9

計量機能（metrological function）

計測マネジメントシステム（**3.5.7**）を定め実施するための執行及び技術に責任をもつ組織機能上の単位.

（**JIS Q 10012**:2011 の **3.6** を変更.）

3.3　活動に関する用語

3.3.1

改善（improvement）

パフォーマンス（**3.7.8**）を向上するための活動.

注記　活動は，繰り返し行われることも，又は一回限りであることもあり得る.

3.3.2

continual improvement

recurring activity to enhance *performance* (3.7.8)

Note 1 to entry: The *process* (3.4.1) of establishing *objectives* (3.7.1) and finding opportunities for *improvement* (3.3.1) is a continual process through the use of *audit findings* (3.13.9) and *audit conclusions* (3.13.10), analysis of *data* (3.8.1), *management* (3.3.3) *reviews* (3.11.2) or other means and generally leads to *corrective action* (3.12.2) or *preventive action* (3.12.1).

Note 2 to entry: This constitutes one of the common terms and core definitions for ISO management system standards given in Annex SL of the Consolidated ISO Supplement to the ISO/IEC Directives, Part 1. The original definition has been modified by adding Note 1 to entry.

3.3.3

management

coordinated activities to direct and control an *or-*

3.3.2

継続的改善（continual improvement）

パフォーマンス（**3.7.8**）を向上するために繰り返し行われる活動.

注記 1　**改善**（**3.3.1**）のための**目標**（**3.7.1**）を設定し，改善の機会を見出す**プロセス**（**3.4.1**）は，**監査所見**（**3.13.9**）及び**監査結論**（**3.13.10**）の利用，**データ**（**3.8.1**）の分析，**マネジメント**（**3.3.3**）**レビュー**（**3.11.2**）又は他の方法を活用した継続的なプロセスであり，一般に**是正処置**（**3.12.2**）又は**予防処置**（**3.12.1**）につながる.

注記 2　この用語及び定義は，ISO/IEC 専門業務用指針―第 1 部：統合版 ISO 補足指針の**附属書 SL** に示された ISO マネジメントシステム規格の共通用語及び中核となる定義の一つを成す．元の定義にない**注記 1** を追加した.

3.3.3

マネジメント，運営管理（management）

組織（**3.2.1**）を指揮し，管理するための調整さ

ganization (3.2.1)

Note 1 to entry: Management can include establishing *policies* (3.5.8) and *objectives* (3.7.1), and *processes* (3.4.1) to achieve these objectives.

Note 2 to entry: The word "management" sometimes refers to people, i.e. a person or group of people with authority and responsibility for the conduct and control of an organization. When "management" is used in this sense, it should always be used with some form of qualifier to avoid confusion with the concept of "management" as a set of activities defined above. For example, "management shall..." is deprecated whereas "*top management* (3.1.1) shall..." is acceptable. Otherwise different words should be adopted to convey the concept when related to people, e.g. managerial or managers.

れた活動.

注記 1　マネジメントには，**方針**（**3.5.8**）及び**目標**（**3.7.1**）の確立，並びにその目標を達成するための**プロセス**（**3.4.1**）が含まれることがある.

注記 2　“マネジメント”という言葉が人を指すことがある．すなわち，組織の指揮及び管理を行うための権限及び責任をもつ個人又はグループを意味することがある．“マネジメント”がこの意味で用いられる場合には，この項で定義した，一連の活動としての“マネジメント”の概念との混同を避けるために，常に何らかの修飾語を付けて用いるのがよい．例えば，“マネジメントは……しなければならない.”は使ってはならないが，“**トップマネジメント**（**3.1.1**）は……しなければならない.”を使うことは許される．このほか，その概念が人に関係することを伝えるために，例えば，“経営者・管理者の”，“管理者”のような別の言葉を用いることが望ましい.

3.3.4

quality management

management (3.3.3) with regard to *quality* (3.6.2)

Note 1 to entry: Quality management can include establishing *quality policies* (3.5.9) and *quality objectives* (3.7.2), and *processes* (3.4.1) to achieve these quality objectives through *quality planning* (3.3.5), *quality assurance* (3.3.6), *quality control* (3.3.7), and *quality improvement* (3.3.8).

3.3.5

quality planning

part of *quality management* (3.3.4) focused on setting *quality objectives* (3.7.2) and specifying necessary operational *processes* (3.4.1), and related resources to achieve the quality objectives

Note 1 to entry: Establishing *quality plans* (3.8.9) can be part of quality planning.

3.3.6

quality assurance

part of *quality management* (3.3.4) focused on

3.3.4

品質マネジメント（quality management）

品質（**3.6.2**）に関する**マネジメント**（**3.3.3**）.

注記　品質マネジメントには,**品質方針**（**3.5.9**）及び**品質目標**（**3.7.2**）の設定，並びに**品質計画**（**3.3.5**），**品質保証**（**3.3.6**），**品質管理**（**3.3.7**）及び**品質改善**（**3.3.8**）を通じてこれらの品質目標を達成するための**プロセス**（**3.4.1**）が含まれ得る.

3.3.5

品質計画（quality planning）

品質目標（**3.7.2**）を設定すること及び必要な運用**プロセス**（**3.4.1**）を規定すること，並びにその品質目標を達成するための関連する資源に焦点を合わせた**品質マネジメント**（**3.3.4**）の一部.

注記　**品質計画書**（**3.8.9**）の作成が，品質計画の一部となる場合がある.

3.3.6

品質保証（quality assurance）

品質要求事項（**3.6.5**）が満たされるという確信

providing confidence that *quality requirements* (3.6.5) will be fulfilled

3.3.7

quality control

part of *quality management* (3.3.4) focused on fulfilling *quality requirements* (3.6.5)

3.3.8

quality improvement

part of *quality management* (3.3.4) focused on increasing the ability to fulfil *quality requirements* (3.6.5)

Note 1 to entry: The quality requirements can be related to any aspect such as *effectiveness* (3.7.11), *efficiency* (3.7.10) or *traceability* (3.6.13).

3.3.9

configuration management

coordinated activities to direct and control *configuration* (3.10.6)

を与えることに焦点を合わせた**品質マネジメント**（**3.3.4**）の一部.

3.3.7

品質管理（quality control）

品質要求事項（**3.6.5**）を満たすことに焦点を合わせた**品質マネジメント**（**3.3.4**）の一部.

3.3.8

品質改善（quality improvement）

品質要求事項（**3.6.5**）を満たす能力を高めることに焦点を合わせた**品質マネジメント**（**3.3.4**）の一部.

注記　品質要求事項は，**有効性**（**3.7.11**），**効率**（**3.7.10**），**トレーサビリティ**（**3.6.13**）などの側面に関連し得る.

3.3.9

コンフィギュレーション管理（configuration management）

コンフィギュレーション（**3.10.6**）を指示し，管理するための調整された活動.

Note 1 to entry: Configuration management generally concentrates on technical and organizational activities that establish and maintain control of a *product* (3.7.6) or *service* (3.7.7) and its *product configuration information* (3.6.8) throughout the life cycle of the product.

[SOURCE: ISO 10007:2003, 3.6, modified — Note 1 to entry has been modified]

3.3.10

change control

<configuration management> activities for control of the *output* (3.7.5) after formal approval of its *product configuration information* (3.6.8)

[SOURCE: ISO 10007:2003, 3.1, modified]

3.3.11

activity

<project management> smallest identified object of work in a *project* (3.4.2)

注記　コンフィギュレーション管理は，一般に，製品のライフサイクルを通して**製品**（**3.7.6**）又は**サービス**（**3.7.7**）及びそれらの**製品コンフィギュレーション情報**（**3.6.8**）の管理を確立及び維持する，技術的活動及び組織的活動に焦点を合わせたものである．

（**ISO　10007**:2003 の **3.6** を変更．**注記 1** を変更した．）

3.3.10

変更管理（change control）

＜コンフィギュレーション管理＞**アウトプット**（**3.7.5**）を，その**製品コンフィギュレーション情報**（**3.6.8**）の正式な承認後に管理するための活動．

（**ISO 10007**:2003 の **3.1** を変更．）

3.3.11

活動（activity）

＜プロジェクトマネジメント＞**プロジェクト**（**3.4.2**）において，明確にされた作業の最小の対象．

[SOURCE: ISO 10006:2003, 3.1, modified]

3.3.12

project management

planning, organizing, *monitoring* (3.11.3), controlling and reporting of all aspects of a *project* (3.4.2), and the motivation of all those involved in it to achieve the project objectives

[SOURCE: ISO 10006:2003, 3.6]

3.3.13

configuration object

object (3.6.1) within a *configuration* (3.10.6) that satisfies an end-use function

[SOURCE: ISO 10007:2003, 3.5, modified]

3.4 Terms related to process

3.4.1

process

set of interrelated or interacting activities that use inputs to deliver an intended result

（**JIS Q 10006**:2004 の **3.1** を変更.）

3.3.12

プロジェクトマネジメント（project management）

プロジェクト（**3.4.2**）の目標を達成するために，プロジェクトの全側面を計画し，組織し，**監視**（**3.11.3**）し，管理し，報告すること，及びプロジェクトに参画する人々全員への動機付けを行うこと.

（**JIS Q 10006**:2004 の **3.6** 参照）

3.3.13

コンフィギュレーション対象（configuration object）

最終使用機能を満たす**コンフィギュレーション**（**3.10.6**）内の**対象**（**3.6.1**）.

（**ISO 10007**:2003 の **3.5** を変更.）

3.4　プロセスに関する用語

3.4.1

プロセス（process）

インプットを使用して意図した結果を生み出す，相互に関連する又は相互に作用する一連の活動.

Note 1 to entry: Whether the "intended result" of a process is called *output* (3.7.5), *product* (3.7.6) or *service* (3.7.7) depends on the context of the reference.
Note 2 to entry: Inputs to a process are generally the outputs of other processes and outputs of a process are generally the inputs to other processes.

Note 3 to entry: Two or more interrelated and interacting processes in series can also be referred to as a process.

Note 4 to entry: Processes in an *organization* (3.2.1) are generally planned and carried out under controlled conditions to add value.
Note 5 to entry: A process where the *conformity* (3.6.11) of the resulting output cannot be readily or economically validated is frequently referred to as a "special process".

Note 6 to entry: This constitutes one of the common terms and core definitions for ISO management system standards given in Annex SL of the

注記 1 プロセスの“意図した結果”を，**アウトプット**（**3.7.5**），**製品**（**3.7.6**）又は**サービス**（**3.7.7**）のいずれと呼ぶかは，それが用いられている文脈による.

注記 2 プロセスへのインプットは，通常，他のプロセスからのアウトプットであり，また，プロセスからのアウトプットは，通常，他のプロセスへのインプットである.

注記 3 連続した二つ又はそれ以上の相互に関連する及び相互に作用するプロセスを，一つのプロセスと呼ぶこともあり得る.

注記 4 **組織**（**3.2.1**）内のプロセスは，価値を付加するために，通常，管理された条件の下で計画され，実行される.

注記 5 結果として得られるアウトプットの**適合**（**3.6.11**）が，容易に又は経済的に確認できないプロセスは，“特殊工程（special process）”と呼ばれることが多い.

注記 6 この用語及び定義は，ISO/IEC 専門業務用指針―第 1 部：統合版 ISO 補足指針の**附属書 SL** に示された ISO

Consolidated ISO Supplement to the ISO/IEC Directives, Part 1. The original definition has been modified to prevent circularity between process and output, and Notes 1 to 5 to entry have been added.

3.4.2

project

unique *process* (3.4.1), consisting of a set of coordinated and controlled activities with start and finish dates, undertaken to achieve an *objective* (3.7.1) conforming to specific *requirements* (3.6.4), including the constraints of time, cost and resources

Note 1 to entry: An individual project can form part of a larger project structure and generally has a defined start and finish date.

Note 2 to entry: In some projects the objectives and scope are updated and the *product* (3.7.6) or *service* (3.7.7) *characteristics* (3.10.1) defined progressively as the project proceeds.

マネジメントシステム規格の共通用語及び中核となる定義の一つを成す．ただし，プロセスの定義とアウトプットの定義との間の循環を防ぐため，元の定義を修正した．また，元の定義にない**注記 1**～**注記 5** を追加した．

3.4.2

プロジェクト（project）

開始日及び終了日をもち，調整され，管理された一連の活動から成り，時間，コスト及び資源の制約を含む特定の**要求事項**（**3.6.4**）に適合する**目標**（**3.7.1**）を達成するために実施される特有の**プロセス**（**3.4.1**）．

注記 1　個別プロジェクトは，より規模の大きいプロジェクトの一部を構成することがあり，通常，定められた開始日及び終了日がある．

注記 2　一部のプロジェクトにおいては，プロジェクトの進行に伴い段階的に，目標及び範囲が更新され，**製品**（**3.7.6**）又は**サービス**（**3.7.7**）の**特性**（**3.10.1**）

Note 3 to entry: The *output* (3.7.5) of a project can be one or several units of product or service.

Note 4 to entry: The project's *organization* (3.2.1) is normally temporary and established for the lifetime of the project.
Note 5 to entry: The complexity of the interactions among project activities is not necessarily related to the project size.

[SOURCE: ISO 10006:2003, 3.5, modified — Notes 1 to 3 have been modified]

3.4.3
quality management system realization

process (3.4.1) of establishing, documenting, implementing, maintaining and continually improving a *quality management system* (3.5.4)

[SOURCE: ISO 10019:2005, 3.1, modified — Notes have been deleted]

が定められていく.

注記 3　プロジェクトの**アウトプット**（**3.7.5**）は，一つの製品又はサービスの場合もあれば，複数の製品又はサービスの場合もある.

注記 4　プロジェクトの**組織**（**3.2.1**）は，通常，一時的なものであり，プロジェクトのライフサイクルに対して設けられる.

注記 5　プロジェクトの活動間における相互作用の複雑さは，プロジェクトの規模に必ずしも関係するとは限らない.

（**JIS Q 10006**:2004 の **3.5** を変更. **参考 1.** ～**参考 3.** を変更した.）

3.4.3

品質マネジメントシステムの実現（quality management system realization）

品質マネジメントシステム（**3.5.4**）を確立し，文書化し，実施し，維持し，継続的に改善する**プロセス**（**3.4.1**）.

（**JIS Q 10019**:2005 の **3.1** を変更. **備考**を削除した.）

3.4.4

competence acquisition

process (3.4.1) of attaining *competence* (3.10.4)

[SOURCE: ISO 10018:2012, 3.2, modified]

3.4.5

procedure

specified way to carry out an activity or a *process* (3.4.1)

Note 1 to entry: Procedures can be documented or not.

3.4.6

outsource (verb)

make an arrangement where an external *organization* (3.2.1) performs part of an organization's function or *process* (3.4.1)

Note 1 to entry: An external organization is outside the scope of the *management system* (3.5.3), although the outsourced function or process is within the scope.

3.4.4

力量の習得（competence acquisition）

力量（**3.10.4**）を身に付ける**プロセス**（**3.4.1**）.

（**ISO 10018**:2012 の **3.2** を変更.）

3.4.5

手順（procedure）

活動又は**プロセス**（**3.4.1**）を実行するために規定された方法.

注記 手順は，文書にすることもあれば，しないこともある.

3.4.6

外部委託する（outsource）（動詞）

ある**組織**（**3.2.1**）の機能又は**プロセス**（**3.4.1**）の一部を外部の組織が実施するという取決めを行う.

注記 1 外部委託した機能又はプロセスは**マネジメントシステム**（**3.5.3**）の適用範囲内にあるが，外部の組織はマネジメントシステムの適用範囲の外にある.

Note 2 to entry: This constitutes one of the common terms and core definitions for ISO management system standards given in Annex SL of the Consolidated ISO Supplement to the ISO/IEC Directives, Part 1.

3.4.7

contract

binding agreement

3.4.8

design and development

set of *processes* (3.4.1) that transform *requirements* (3.6.4) for an *object* (3.6.1) into more detailed requirements for that object

Note 1 to entry: The requirements forming input to design and development are often the result of research and can be expressed in a broader, more general sense than the requirements forming the *output* (3.7.5) of design and development. The requirements are generally defined in terms of *characteristics* (3.10.1). In a *project* (3.4.2) there can be several design and development stages.

注記 2　この用語及び定義は，ISO/IEC 専門業務用指針—第 1 部：統合版 ISO 補足指針の**附属書 SL** に示された ISO マネジメントシステム規格の共通用語及び中核となる定義の一つを成す．

3.4.7

契約（contract）

拘束力のある取決め．

3.4.8

設計・開発（design and development）

対象（**3.6.1**）に対する**要求事項**（**3.6.4**）を，その対象に対するより詳細な要求事項に変換する一連の**プロセス**（**3.4.1**）．

注記 1　設計・開発へのインプットとなる要求事項は，調査・研究の結果であることが多く，また，設計・開発からの**アウトプット**（**3.7.5**）となる要求事項よりも広範で，一般的な意味で表現されることがある．要求事項は，通常，**特性**（**3.10.1**）を用いて定義される．**プロジェクト**（**3.4.2**）には，複数の設計・

Note 2 to entry: In English the words "design" and "development" and the term "design and development" are sometimes used synonymously and sometimes used to define different stages of the overall design and development. In French the words "conception" and "développement" and the term "conception et développement" are sometimes used synonymously and sometimes used to define different stages of the overall design and development.
Note 3 to entry: A qualifier can be applied to indicate the nature of what is being designed and developed (e.g. *product* (3.7.6) design and development, *service* (3.7.7) design and development or process design and development).

3.5 Terms related to system

3.5.1

system

set of interrelated or interacting elements

開発段階が存在することがある.

注記 2　“設計”,“開発”及び“設計・開発”という言葉は,あるときは同じ意味で使われ,あるときには設計・開発全体の異なる段階を定義するために使われる.

注記 3　設計・開発されるものの性格を示すために,修飾語が用いられることがある[**例**　**製品**(**3.7.6**)の設計・開発,**サービス**(**3.7.7**)の設計・開発又はプロセスの設計・開発].

3.5　システムに関する用語

3.5.1

システム(system)

相互に関連する又は相互に作用する要素の集まり.

3.5.2

infrastructure

<organization> *system* (3.5.1) of facilities, equipment and *services* (3.7.7) needed for the operation of an *organization* (3.2.1)

3.5.3

management system

set of interrelated or interacting elements of an *organization* (3.2.1) to establish *policies* (3.5.8) and *objectives* (3.7.1), and *processes* (3.4.1) to achieve those objectives

Note 1 to entry: A management system can address a single discipline or several disciplines, e.g. *quality management* (3.3.4), financial management or environmental management.

Note 2 to entry: The management system elements establish the organization's structure, roles and responsibilities, planning, operation, policies, practices, rules, beliefs, objectives and processes to achieve those objectives.

Note 3 to entry: The scope of a management sys-

3.5.2

インフラストラクチャ（infrastructure）

＜組織＞**組織**（**3.2.1**）の運営のために必要な施設，設備及び**サービス**（**3.7.7**）に関する**システム**（**3.5.1**）.

3.5.3

マネジメントシステム（management system）

方針（**3.5.8**）及び**目標**（**3.7.1**），並びにその目標を達成するための**プロセス**（**3.4.1**）を確立するための，相互に関連する又は相互に作用する，**組織**（**3.2.1**）の一連の要素.

注記 1　一つのマネジメントシステムは，例えば，**品質マネジメント**（**3.3.4**），財務マネジメント，環境マネジメントなど，単一又は複数の分野を取り扱うことができる.

注記 2　マネジメントシステムの要素は，目標を達成するための，組織の構造，役割及び責任，計画，運用，方針，慣行，規則，信条，目標，並びにプロセスを確立する.

注記 3　マネジメントシステムの適用範囲とし

tem can include the whole of the organization, specific and identified functions of the organization, specific and identified sections of the organization, or one or more functions across a group of organizations.

Note 4 to entry: This constitutes one of the common terms and core definitions for ISO management system standards given in Annex SL of the Consolidated ISO Supplement to the ISO/IEC Directives, Part 1. The original definition has been modified by modifying Notes 1 to 3 to entry.

3.5.4

quality management system

part of a *management system* (3.5.3) with regard to *quality* (3.6.2)

3.5.5

work environment

set of conditions under which work is performed

Note 1 to entry: Conditions can include physical, social, psychological and environmental factors

ては，組織全体，組織内の固有で特定された機能，組織内の固有で特定された部門，複数の組織の集まりを横断する一つ又は複数の機能，などがあり得る．

注記 4　この用語及び定義は，ISO/IEC 専門業務用指針—第 1 部：統合版 ISO 補足指針の**附属書 SL** に示された ISO マネジメントシステム規格の共通用語及び中核となる定義の一つを成す．元の定義の**注記 1**～**注記 3** を変更した．

3.5.4

品質マネジメントシステム（quality management system）

品質（**3.6.2**）に関する，**マネジメントシステム**（**3.5.3**）の一部．

3.5.5

作業環境（work environment）

作業が行われる場の条件の集まり．

注記　条件には，物理的，社会的，心理的及び環境的要因を含み得る（例えば，温度，

(such as temperature, lighting, recognition schemes, occupational stress, ergonomics and atmospheric composition).

3.5.6

metrological confirmation

set of operations required to ensure that *measuring equipment* (3.11.6) conforms to the *requirements* (3.6.4) for its intended use

Note 1 to entry: Metrological confirmation generally includes calibration or *verification* (3.8.12), any necessary adjustment or *repair* (3.12.9), and subsequent recalibration, comparison with the metrological requirements for the intended use of the equipment, as well as any required sealing and labelling.

Note 2 to entry: Metrological confirmation is not achieved until and unless the fitness of the measuring equipment for the intended use has been demonstrated and documented.

Note 3 to entry: The requirements for intended use include such considerations as range, resolution and maximum permissible errors.

照明，表彰制度，業務上のストレス，人間工学的側面，大気成分）.

3.5.6

計量確認（metrological confirmation）

測定機器（**3.11.6**）が，その意図した用途の**要求事項**（**3.6.4**）に適合していることを確認するために必要な一連の操作.

注記 1　計量確認は，通常，校正又は**検証**（**3.8.12**），必要な調整又は**修理**（**3.12.9**）及びその後の再校正，機器の意図した用途に関する計量要求事項との比較，並びに必要な一切の封印及びラベル表示を含む.

注記 2　計量確認は，測定機器が意図した用途に適していることが立証されて文書化されるまで，又は立証されて文書化されない限り，達成されない.

注記 3　意図した用途に関する要求事項には，測定範囲，分解能，最大許容誤差などの考慮事項を含む.

Note 4 to entry: Metrological requirements are usually distinct from, and are not specified in, *product* (3.7.6) requirements.

[SOURCE: ISO 10012:2003, 3.5, modified — Note 1 to entry has been modified]

3.5.7

measurement management system

set of interrelated or interacting elements necessary to achieve *metrological confirmation* (3.5.6) and control of *measurement processes* (3.11.5)

[SOURCE: ISO 10012:2003, 3.1, modified]

3.5.8

policy

<organization> intentions and direction of an *organization* (3.2.1) as formally expressed by its *top management* (3.1.1)

Note 1 to entry: This constitutes one of the common terms and core definitions for ISO manage-

注記 4　計量要求事項は，通常，**製品**（**3.7.6**）要求事項とは別のものであり，製品要求事項には規定されない．

（**JIS Q 10012**:2011 の **3.5** を変更．**注記 1** を変更した．）

3.5.7

計測マネジメントシステム（measurement management system）

計量確認（**3.5.6**）及び**測定プロセス**（**3.11.5**）の管理の達成に必要な，相互に関係する又は相互に作用する一連の要素．

（**JIS Q 10012**:2011 の **3.1** を変更．）

3.5.8

方針（policy）

<組織>**トップマネジメント**（**3.1.1**）によって正式に表明された**組織**（**3.2.1**）の意図及び方向付け．

注記　この用語及び定義は，ISO/IEC 専門業務用指針—第 1 部：統合版 ISO 補足指

ment system standards given in Annex SL of the Consolidated ISO Supplement to the ISO/IEC Directives, Part 1.

3.5.9

quality policy

policy (3.5.8) related to *quality* (3.6.2)

Note 1 to entry: Generally the quality policy is consistent with the overall policy of the *organization* (3.2.1), can be aligned with the organization's *vision* (3.5.10) and *mission* (3.5.11) and provides a framework for the setting of *quality objectives* (3.7.2).

Note 2 to entry: Quality management principles presented in this International Standard can form a basis for the establishment of a quality policy.

3.5.10

vision

<organization> aspiration of what an *organization* (3.2.1) would like to become as expressed by *top management* (3.1.1)

針の**附属書 SL** に示された ISO マネジメントシステム規格の共通用語及び中核となる定義の一つを成す.

3.5.9

品質方針（quality policy）

品質（**3.6.2**）に関する**方針**（**3.5.8**）.

注記 1　一般に品質方針は，**組織**（**3.2.1**）の全体的な方針と整合しており，組織の**ビジョン**（**3.5.10**）及び**使命**（**3.5.11**）と密接に関連付けることができ，**品質目標**（**3.7.2**）を設定するための枠組みを提供する.

注記 2　この規格に記載した品質マネジメントの原則は，品質方針を設定するための基礎となり得る.

3.5.10

ビジョン（vision）

<組織>**トップマネジメント**（**3.1.1**）によって表明された，**組織**（**3.2.1**）がどのようになりたいかについての願望.

3.5.11

mission

<organization> *organization's* (3.2.1) purpose for existing as expressed by *top management* (3.1.1)

3.5.12

strategy

plan to achieve a long-term or overall *objective* (3.7.1)

3.6 Terms related to requirement

3.6.1

object

entity

item

anything perceivable or conceivable

EXAMPLE *Product* (3.7.6), *service* (3.7.7), *process* (3.4.1), person, *organization* (3.2.1), *system* (3.5.1), resource.

Note 1 to entry: Objects can be material (e.g. an engine, a sheet of paper, a diamond), non-material (e.g. conversion ratio, a project plan) or imag-

3.5.11

使命（mission）

＜組織＞**トップマネジメント**（**3.1.1**）によって表明された，**組織**（**3.2.1**）の存在目的.

3.5.12

戦略（strategy）

長期的又は全体的な**目標**（**3.7.1**）を達成するための計画.

3.6　要求事項に関する用語

3.6.1

対象（object），**実体**（entity），**項目**（item）

認識できるもの又は考えられるもの全て.

例　**製品**（**3.7.6**），**サービス**（**3.7.7**），**プロセス**（**3.4.1**），人，**組織**（**3.2.1**），**システム**（**3.5.1**），資源

注記　対象は，物質的なもの（**例**　エンジン，一枚の紙，ダイヤモンド），非物質的なもの（**例**　変換率，プロジェクト計画），

ined (e.g. the future state of the organization).

[SOURCE: ISO 1087-1:2000, 3.1.1, modified]

3.6.2

quality

degree to which a set of inherent *characteristics* (3.10.1) of an *object* (3.6.1) fulfils *requirements* (3.6.4)

Note 1 to entry: The term "quality" can be used with adjectives such as poor, good or excellent.

Note 2 to entry: "Inherent" , as opposed to "assigned" , means existing in the *object* (3.6.1).

3.6.3

grade

category or rank given to different *requirements* (3.6.4) for an *object* (3.6.1) having the same functional use

又は想像上のもの（**例**　組織の将来の状態）の場合がある.

（**ISO 1087-1**:2000 の **3.1.1** を変更.）

3.6.2

品質（quality）

対象（**3.6.1**）に本来備わっている**特性**（**3.10.1**）の集まりが，**要求事項**（**3.6.4**）を満たす程度.

注記 1　“品質”という用語は，悪い，良い，優れたなどの形容詞とともに使われることがある.

注記 2　“本来備わっている”とは，“付与された”とは異なり，**対象**（**3.6.1**）の中に存在していることを意味する.

3.6.3

等級（grade）

同一の用途をもつ**対象**（**3.6.1**）の，異なる**要求事項**（**3.6.4**）に対して与えられる区分又はランク.

EXAMPLE Class of airline ticket and category of hotel in a hotel brochure.

Note 1 to entry: When establishing a *quality requirement* (3.6.5), the grade is generally specified.

3.6.4

requirement

need or expectation that is stated, generally implied or obligatory

Note 1 to entry: "Generally implied" means that it is custom or common practice for the *organization* (3.2.1) and *interested parties* (3.2.3) that the need or expectation under consideration is implied.

Note 2 to entry: A specified requirement is one that is stated, for example in *documented information* (3.8.6).

Note 3 to entry: A qualifier can be used to denote a specific type of requirement, e.g. *product* (3.7.6) requirement, *quality management* (3.3.4) requirement, *customer* (3.2.4) requirement, *quality re-*

例 航空券のクラス，ホテルの案内書に示されるホテルの区分

注記 **品質要求事項**（**3.6.5**）を設定する場合，通常，その等級を規定する．

3.6.4

要求事項（requirement）

明示されている，通常暗黙のうちに了解されている又は義務として要求されている，ニーズ又は期待．

注記 1 “通常暗黙のうちに了解されている” とは，対象となるニーズ又は期待が暗黙のうちに了解されていることが，**組織**（**3.2.1**）及び**利害関係者**（**3.2.3**）にとって，慣習又は慣行であることを意味する．

注記 2 規定要求事項とは，例えば，**文書化した情報**（**3.8.6**）の中で明示されている要求事項をいう．

注記 3 特定の種類の要求事項であることを示すために，修飾語を用いることがある．

例 **製品**（**3.7.6**）要求事項，**品質マネジメント**（**3.3.4**）要求事項，

quirement (3.6.5).

Note 4 to entry: Requirements can be generated by different interested parties or by the organization itself.

Note 5 to entry: It can be necessary for achieving high *customer satisfaction* (3.9.2) to fulfil an expectation of a customer even if it is neither stated nor generally implied or obligatory.

Note 6 to entry: This constitutes one of the common terms and core definitions for ISO management system standards given in Annex SL of the Consolidated ISO Supplement to the ISO/IEC Directives, Part 1. The original definition has been modified by adding Notes 3 to 5 to entry.

3.6.5

quality requirement

requirement (3.6.4) related to *quality* (3.6.2)

顧客（3.2.4）要求事項，**品質要求事項（3.6.5）**

注記 4　要求事項は，異なる利害関係者又は組織自身から出されることがある．

注記 5　顧客の期待が明示されていない，暗黙のうちに了解されていない又は義務として要求されていない場合でも，高い**顧客満足（3.9.2）**を達成するために顧客の期待を満たすことが必要なことがある．

注記 6　この用語及び定義は，ISO/IEC 専門業務用指針—第 1 部：統合版 ISO 補足指針の**附属書 SL** に示された ISO マネジメントシステム規格の共通用語及び中核となる定義の一つを成す．元の定義にない**注記 3** ～**注記 5** を追加した．

3.6.5

品質要求事項（quality requirement）

品質（3.6.2）に関する**要求事項（3.6.4）**．

3.6.6

statutory requirement

obligatory *requirement* (3.6.4) specified by a legislative body

3.6.7

regulatory requirement

obligatory *requirement* (3.6.4) specified by an authority mandated by a legislative body

3.6.8

product configuration information

requirement (3.6.4) or other information for *product* (3.7.6) design, realization, *verification* (3.8.12), operation and support

[SOURCE: ISO 10007:2003, 3.9, modified]

3.6.9

nonconformity

non-fulfilment of a *requirement* (3.6.4)

Note 1 to entry: This constitutes one of the com-

3.6.6

法令要求事項（statutory requirement）

立法機関によって規定された，必須の**要求事項**（**3.6.4**）.

3.6.7

規制要求事項（regulatory requirement）

立法機関から委任された当局によって規定された，必須の**要求事項**（**3.6.4**）.

3.6.8

製品コンフィギュレーション情報（product configuration information）

製品（**3.7.6**）の設計，実現，**検証**（**3.8.12**），運用及びサポートに関する**要求事項**（**3.6.4**）又はその他の情報.

（**ISO 10007**:2003 の **3.9** を変更.）

3.6.9

不適合（nonconformity）

要求事項（**3.6.4**）を満たしていないこと.

注記　この用語及び定義は，ISO/IEC 専門業

mon terms and core definitions for ISO management system standards given in Annex SL of the Consolidated ISO Supplement to the ISO/IEC Directives, Part 1.

3.6.10

defect

nonconformity (3.6.9) related to an intended or specified use

Note 1 to entry: The distinction between the concepts defect and nonconformity is important as it has legal connotations, particularly those associated with *product* (3.7.6) and *service* (3.7.7) liability issues.

Note 2 to entry: The intended use as intended by the *customer* (3.2.4) can be affected by the nature of the *information* (3.8.2), such as operating or maintenance instructions, provided by the *provider* (3.2.5).

務用指針—第1部：統合版ISO補足指針の**附属書SL**に示されたISOマネジメントシステム規格の共通用語及び中核となる定義の一つを成す.

3.6.10

欠陥（defect）

意図された用途又は規定された用途に関する**不適合**（**3.6.9**）.

注記1 欠陥と不適合という概念の区別は，特に**製品**（**3.7.6**）及び**サービス**（**3.7.7**）の製造物責任問題に関連している場合には,法的意味をもつので重要である.

注記2 **顧客**（**3.2.4**）によって意図される用途は，**提供者**（**3.2.5**）から提供される**情報**（**3.8.2**）の性質によって影響を受けることがある．これらの情報には，例えば，取扱説明書，メンテナンス説明書などがある.

3.6.11

conformity

fulfilment of a *requirement* (3.6.4)

Note 1 to entry: In English the word "conformance" is synonymous but deprecated. In French the word "compliance" is synonymous but deprecated.

Note 2 to entry: This constitutes one of the common terms and core definitions for ISO management system standards given in Annex SL of the Consolidated ISO Supplement to the ISO/IEC Directives, Part 1. The original definition has been modified by adding Note 1 to entry.

3.6.12

capability

ability of an *object* (3.6.1) to realize an *output* (3.7.5) that will fulfil the *requirements* (3.6.4) for that output

Note 1 to entry: *Process* (3.4.1) capability terms in the field of statistics are defined in ISO 3534-2.

3.6.11

適合（conformity）

要求事項（**3.6.4**）を満たしていること．

注記 1　対応国際規格の注記では，英語及びフランス語の同義語について説明しているが，この規格では不要であり，削除した．

注記 2　この用語及び定義は，ISO/IEC 専門業務用指針―第 1 部：統合版 ISO 補足指針の**附属書 SL** に示された ISO マネジメントシステム規格の共通用語及び中核となる定義の一つを成す．元の定義にない**注記 1** を追加した．

3.6.12

実現能力（capability）

要求事項（**3.6.4**）を満たす**アウトプット**（**3.7.5**）を実現する，**対象**（**3.6.1**）の能力．

注記　統計の分野における工程能力の用語は，**JIS Z 8101-2** に定義されている．

3.6.13

traceability

ability to trace the history, application or location of an *object* (3.6.1)

Note 1 to entry: When considering a *product* (3.7.6) or a *service* (3.7.7), traceability can relate to:

— the origin of materials and parts;

— the processing history;

— the distribution and location of the product or service after delivery.

Note 2 to entry: In the field of metrology, the definition in ISO/IEC Guide 99 is the accepted definition.

3.6.14

dependability

ability

to perform as and when required

[SOURCE: IEC 60050-192, modified — Notes have been deleted]

3.6.13

トレーサビリティ（traceability）

対象（**3.6.1**）の履歴，適用又は所在を追跡できること．

注記 1　**製品**（**3.7.6**）又は**サービス**（**3.7.7**）に関しては，トレーサビリティは，次のようなものに関連することがある．

— 材料及び部品の源

— 処理の履歴

— 製品又はサービスの提供後の分布及び所在

注記 2　計量計測の分野においては，**ISO/IEC Guide 99** に記載する定義が受け入れられている．

3.6.14

ディペンダビリティ（dependability）

求められたとおりに，かつ，求められたときに，機能する能力．

（**IEC 60050-192** を変更．**注記**を削除した．）

3.6.15

innovation

new or changed *object* (3.6.1) realizing or redistributing value

Note 1 to entry: Activities resulting in innovation are generally managed.

Note 2 to entry: Innovation is generally significant in its effect.

3.7 Terms related to result

3.7.1

objective

result to be achieved

Note 1 to entry: An objective can be strategic, tactical, or operational.

Note 2 to entry: Objectives can relate to different disciplines (such as financial, health and safety, and environmental objectives) and can apply at different levels (such as strategic, *organization* (3.2.1)-wide, *project* (3.4.2), *product* (3.7.6) and *process* (3.4.1)).

Note 3 to entry: An objective can be expressed in

3.6.15

革新（innovation）

価値を実現する又は再配分する，新しい又は変更された**対象**（**3.6.1**）．

注記 1　革新を結果として生む活動は，一般に，マネジメントされている．

注記 2　革新は，一般に，その影響が大きい．

3.7　結果に関する用語

3.7.1

目標（objective）

達成すべき結果．

注記 1　目標は，戦略的，戦術的又は運用的であり得る．

注記 2　目標は，様々な領域（例えば，財務，安全衛生，環境目標）に関連し得るものであり，様々な階層［例えば，戦略的レベル，**組織**（**3.2.1**）全体，**プロジェクト**（**3.4.2**）単位，**製品**（**3.7.6**）ごと，**プロセス**（**3.4.1**）ごと］で適用できる．

注記 3　目標は，例えば，意図する成果，目的，

other ways, e.g. as an intended outcome, a purpose, an operational criterion, as a *quality objective* (3.7.2) or by the use of other words with similar meaning (e.g. aim, goal, or target).

Note 4 to entry: In the context of *quality management systems* (3.5.4) *quality objectives* (3.7.2) are set by the *organization* (3.2.1), consistent with the *quality policy* (3.5.9), to achieve specific results.

Note 5 to entry: This constitutes one of the common terms and core definitions for ISO management system standards given in Annex SL of the Consolidated ISO Supplement to the ISO/IEC Directives, Part 1. The original definition has been modified by modifying Note 2 to entry.

3.7.2

quality objective

objective (3.7.1) related to *quality* (3.6.2)

Note 1 to entry: Quality objectives are generally based on the *organization's* (3.2.1) *quality policy*

運用基準など，別の形で表現することもできる．また，**品質目標**（**3.7.2**）という表現の仕方もある．又は，同じような意味をもつ別の言葉［**例**　狙い（aim），到達点（goal），標的（target）］で表すこともできる．

注記 4　**品質マネジメントシステム**（**3.5.4**）の場合，**組織**（**3.2.1**）は，特定の結果を達成するため，**品質方針**（**3.5.9**）と整合のとれた**品質目標**（**3.7.2**）を設定する．

注記 5　この用語及び定義は，ISO/IEC 専門業務用指針—第 1 部：統合版 ISO 補足指針の**附属書 SL** に示された ISO マネジメントシステム規格の共通用語及び中核となる定義の一つを成す．元の定義の**注記 2** を変更した．

3.7.2

品質目標（quality objective）

品質（**3.6.2**）に関する**目標**（**3.7.1**）．

注記 1　品質目標は，通常，**組織**（**3.2.1**）の**品質方針**（**3.5.9**）に基づいている．

(3.5.9).

Note 2 to entry: Quality objectives are generally specified for relevant functions, levels and *processes* (3.4.1) in the *organization* (3.2.1).

3.7.3

success

<organization> achievement of an *objective* (3.7.1)

Note 1 to entry: The success of an *organization* (3.2.1) emphasizes the need for a balance between its economic or financial interests and the needs of its *interested parties* (3.2.3), such as *customers* (3.2.4), users, investors/shareholders (owners), people in the organization, *providers* (3.2.5), partners, interest groups and communities.

3.7.4

sustained success

<organization> *success* (3.7.3) over a period of time

Note 1 to entry: Sustained success emphasizes the need for a balance between economic-financial

注記 2　品質目標は，通常，**組織**（**3.2.1**）内の関連する機能，階層及び**プロセス**（**3.4.1**）に対して規定される．

3.7.3

成功（success）

＜組織＞**目標**（**3.7.1**）の達成．

注記　**組織**（**3.2.1**）の成功については，組織の経済的又は財務的利益と，例えば，**顧客**（**3.2.4**），利用者，投資者・株主（所有者），組織内の人々，**提供者**（**3.2.5**），パートナ，利益団体，地域社会などの**利害関係者**（**3.2.3**）のニーズとのバランスをとることの必要性が重視される．

3.7.4

持続的成功（sustained success）

＜組織＞長期にわたる**成功**（**3.7.3**）．

注記 1　持続的成功については，**組織**（**3.2.1**）の経済的・財務的利益と，社会的及び

interests of an *organization* (3.2.1) and those of the social and ecological environment.

Note 2 to entry: Sustained success relates to the *interested parties* (3.2.3) of an organization, such as *customers* (3.2.4), owners, people in an organization, *providers* (3.2.5), bankers, unions, partners or society.

3.7.5

output

result of a *process* (3.4.1)

Note 1 to entry: Whether an output of the *organization* (3.2.1) is a *product* (3.7.6) or a *service* (3.7.7) depends on the preponderance of the *characteristics* (3.10.1) involved, e.g. a painting for sale in a gallery is a product whereas supply of a commissioned painting is a service, a hamburger bought in a retail store is a product whereas receiving an order and serving a hamburger ordered in a restaurant is part of a service.

生態学的環境の利益とのバランスをとることの必要性が重視される.

注記 2　持続的成功は, 例えば, **顧客**（**3.2.4**）, 所有者, 組織内の人々, **提供者**（**3.2.5**）, 銀行家, 組合, パートナ, 社会など, 組織の**利害関係者**（**3.2.3**）に関係する.

3.7.5

アウトプット（output）

プロセス（**3.4.1**）の結果.

注記　**組織**（**3.2.1**）のアウトプットが**製品**（**3.7.6**）又は**サービス**（**3.7.7**）のいずれであるかは, アウトプットがもっている**特性**（**3.10.1**）のうちのどれが優位かということに左右される.

例　画廊で売出し中の絵画は製品であるのに対して, 委託された絵画を提供することはサービスである. 小売店で購入したハンバーガーは製品であるのに対して, レストランでハンバーガーの注文を受け, 提供することはサービスの一部で

3.7.6

product

output (3.7.5) of an *organization* (3.2.1) that can be produced without any transaction taking place between the organization and the *customer* (3.2.4)

Note 1 to entry: Production of a product is achieved without any transaction necessarily taking place between *provider* (3.2.5) and customer, but can often involve this *service* (3.7.7) element upon its delivery to the customer.

Note 2 to entry: The dominant element of a product is that it is generally tangible.

Note 3 to entry: Hardware is tangible and its amount is a countable *characteristic* (3.10.1) (e.g. tyres). Processed materials are tangible and their amount is a continuous characteristic (e.g. fuel and soft drinks). Hardware and processed materials are often referred to as goods. Software consists of *information* (3.8.2) regardless of delivery medium (e.g. computer programme, mobile phone

ある.

3.7.6

製品（product）

組織（**3.2.1**）と**顧客**（**3.2.4**）との間の処理・行為なしに生み出され得る，組織の**アウトプット**（**3.7.5**）.

注記 1　製品の製造は，**提供者**（**3.2.5**）と顧客との間で行われる処理・行為なしでも達成されるが，顧客への引き渡しにおいては，提供者と顧客との間で行われる処理・行為のような**サービス**（**3.7.7**）要素を伴う場合が多い.

注記 2　製品の主要な要素は，一般にそれが有形であることである.

注記 3　ハードウェアは，有形で，その量は数えることができる**特性**（**3.10.1**）をもつ（**例**　タイヤ）．素材製品は，有形で，その量は連続的な特性をもつ（**例**　燃料，清涼飲料水）．ハードウェア及び素材製品は，品物と呼ぶ場合が多い．ソフトウェアは，提供媒体にかかわらず，**情報**（**3.8.2**）から構成される（**例**

app, instruction manual, dictionary content, musical composition copyright, driver's license).

3.7.7

service

output (3.7.5) of an *organization* (3.2.1) with at least one activity necessarily performed between the organization and the *customer* (3.2.4)

Note 1 to entry: The dominant elements of a service are generally intangible.

Note 2 to entry: Service often involves activities at the interface with the customer to establish customer *requirements* (3.6.4) as well as upon delivery of the service and can involve a continuing relationship such as banks, accountancies or public organizations, e.g. schools or hospitals.

Note 3 to entry: Provision of a service can involve, for example, the following:

— an activity performed on a customer-supplied

コンピュータプログラム，携帯電話のアプリケーション，指示マニュアル，辞書コンテンツ，音楽の作曲著作権，運転免許）.

3.7.7

サービス（service）

組織（**3.2.1**）と**顧客**（**3.2.4**）との間で必ず実行される，少なくとも一つの活動を伴う組織の**アウトプット**（**3.7.5**）.

注記 1　サービスの主要な要素は，一般にそれが無形であることである.

注記 2　サービスは，サービスを提供するときに活動を伴うだけでなく，顧客とのインターフェースにおける，顧客**要求事項**（**3.6.4**）を設定するための活動を伴うことが多く，また，銀行，会計事務所，公的機関（**例**　学校，病院）などのように継続的な関係を伴う場合が多い.

注記 3　サービスの提供には，例えば，次のものがあり得る.

— 顧客支給の有形の**製品**（**3.7.6**）（**例**

tangible *product* (3.7.6) (e.g. a car to be repaired);

— an activity performed on a customer-supplied intangible product (e.g. the income statement needed to prepare a tax return);

— the delivery of an intangible product (e.g. the delivery of *information* (3.8.2) in the context of knowledge transmission);

— the creation of ambience for the customer (e.g. in hotels and restaurants);

Note 4 to entry: A service is generally experienced by the customer.

3.7.8

performance

measurable result

Note 1 to entry: Performance can relate either to quantitative or qualitative findings.

Note 2 to entry: Performance can relate to the *management* (3.3.3) of *activities* (3.3.11), *processes* (3.4.1), *products* (3.7.6), *services* (3.7.7), *systems* (3.5.1) or *organizations* (3.2.1).

修理される車）に対して行う活動

— 顧客支給の無形の製品（**例**　納税申告に必要な収支情報）に対して行う活動
— 無形の製品の提供［**例**　知識伝達という意味での**情報**(**3.8.2**)提供］

— 顧客のための雰囲気作り（**例**　ホテル内，レストラン内）

注記 4　サービスは，一般に，顧客によって経験される．

3.7.8

パフォーマンス（performance）

測定可能な結果．

注記 1　パフォーマンスは，定量的又は定性的な所見のいずれにも関連し得る．

注記 2　パフォーマンスは，**活動**（**3.3.11**），**プロセス**（**3.4.1**），**製品**（**3.7.6**），**サービス**（**3.7.7**），**システム**（**3.5.1**）又は**組織**（**3.2.1**）の**運営管理**（**3.3.3**）に関連し得る．

Note 3 to entry: This constitutes one of the common terms and core definitions for ISO management system standards given in Annex SL of the Consolidated ISO Supplement to the ISO/IEC Directives, Part 1. The original definition has been modified by modifying Note 2 to entry.

3.7.9

risk

effect of uncertainty

Note 1 to entry: An effect is a deviation from the expected — positive or negative.

Note 2 to entry: Uncertainty is the state, even partial, of deficiency of *information* (3.8.2) related to, understanding or knowledge of, an event, its consequence, or likelihood.

Note 3 to entry: Risk is often characterized by reference to potential events (as defined in ISO Guide 73:2009, 3.5.1.3) and consequences (as defined in ISO Guide 73:2009, 3.6.1.3), or a combination of these.

注記 3　この用語及び定義は，ISO/IEC 専門業務用指針―第 1 部：統合版 ISO 補足指針の**附属書 SL** に示された ISO マネジメントシステム規格の共通用語及び中核となる定義の一つを成す．元の定義の**注記 2** を変更した．

3.7.9

リスク（risk）

不確かさの影響．

注記 1　影響とは，期待されていることから，好ましい方向又は好ましくない方向にかい（乖）離することをいう．

注記 2　不確かさとは，事象，その結果又はその起こりやすさに関する，**情報**（**3.8.2**），理解又は知識に，たとえ部分的にでも不備がある状態をいう．

注記 3　リスクは，起こり得る事象（**JIS Q 0073**:2010 の **3.5.1.3** の定義を参照．）及び結果（**JIS Q 0073**:2010 の **3.6.1.3** の定義を参照．），又はこれらの組合せについて述べることによって，その特徴を示すことが多い．

Note 4 to entry: Risk is often expressed in terms of a combination of the consequences of an event (including changes in circumstances) and the associated likelihood (as defined in ISO Guide 73:2009, 3.6.1.1) of occurrence.

Note 5 to entry: The word "risk" is sometimes used when there is the possibility of only negative consequences.

Note 6 to entry: This constitutes one of the common terms and core definitions for ISO management system standards given in Annex SL of the Consolidated ISO Supplement to the ISO/IEC Directives, Part 1. The original definition has been modified by adding Note 5 to entry.

3.7.10

efficiency

relationship between the result achieved and the resources used

3.7.11

effectiveness

extent to which planned activities are realized and planned results are achieved

注記 4　リスクは，ある事象（その周辺状況の変化を含む.）の結果とその発生の起こりやすさ（**JIS Q 0073**:2010 の **3.6.1.1** の定義を参照.）との組合せとして表現されることが多い.

注記 5　“リスク”という言葉は，好ましくない結果にしかならない可能性の場合に使われることがある.

注記 6　この用語及び定義は，ISO/IEC 専門業務用指針—第 1 部：統合版 ISO 補足指針の**附属書 SL** に示された ISO マネジメントシステム規格の共通用語及び中核となる定義の一つを成す．元の定義にない**注記 5** を追加した.

3.7.10

効率（efficiency）

達成された結果と使用された資源との関係.

3.7.11

有効性（effectiveness）

計画した活動を実行し，計画した結果を達成した程度.

Note 1 to entry: This constitutes one of the common terms and core definitions for ISO management system standards given in Annex SL of the Consolidated ISO Supplement to the ISO/IEC Directives, Part 1. The original definition has been modified by adding "are" before "achieved".

3.8 Terms related to data, information and document

3.8.1

data

facts about an *object* (3.6.1)

3.8.2

information

meaningful *data* (3.8.1)

3.8.3

objective evidence

data (3.8.1) supporting the existence or verity of something

Note 1 to entry: Objective evidence can be obtained through observation, *measurement* (3.11.4),

注記 この用語及び定義は，ISO/IEC 専門業務用指針―第 1 部：統合版 ISO 補足指針の**附属書 SL** に示された ISO マネジメントシステム規格の共通用語及び中核となる定義の一つを成す.

3.8 データ，情報及び文書に関する用語

3.8.1

データ（data）

対象（**3.6.1**）に関する事実.

3.8.2

情報（information）

意味のある**データ**（**3.8.1**）.

3.8.3

客観的証拠（objective evidence）

あるものの存在又は真実を裏付ける**データ**（**3.8.1**）.

注記 1 客観的証拠は，観察，**測定**（**3.11.4**），**試験**（**3.11.8**），又はその他の手段に

test (3.11.8), or by other means.

Note 2 to entry: Objective evidence for the purpose of *audit* (3.13.1) generally consists of *records* (3.8.10), statements of fact or other *information* (3.8.2) which are relevant to the *audit criteria* (3.13.7) and verifiable.

3.8.4

information system

<quality management system> network of communication channels used within an *organization* (3.2.1)

3.8.5

document

information (3.8.2) and the medium on which it is contained

EXAMPLE *Record* (3.8.10), *specification* (3.8.7), procedure document, drawing, report, standard.

Note 1 to entry: The medium can be paper, magnetic, electronic or optical computer disc, photograph or master sample, or combination thereof.

よって得ることができる.

注記 2　**監査**（**3.13.1**）のための客観的証拠は，一般に，**監査基準**（**3.13.7**）に関連し，かつ，検証できる，**記録**（**3.8.10**），事実の記述又はその他の**情報**（**3.8.2**）から成る.

3.8.4

情報システム（information system）

<品質マネジメントシステム>**組織**（**3.2.1**）内で用いられる，コミュニケーション経路のネットワーク.

3.8.5

文書（document）

情報（**3.8.2**）及びそれが含まれている媒体.

例　**記録**（**3.8.10**），**仕様書**（**3.8.7**），手順を記した文書，図面，報告書，規格

注記 1　媒体としては，紙，磁気，電子式若しくは光学式コンピュータディスク，写真若しくはマスターサンプル，又はこ

Note 2 to entry: A set of documents, for example specifications and records, is frequently called "documentation".

Note 3 to entry: Some *requirements* (3.6.4) (e.g. the requirement to be readable) relate to all types of documents. However there can be different requirements for specifications (e.g. the requirement to be revision controlled) and for records (e.g. the requirement to be retrievable).

3.8.6

documented information

information (3.8.2) required to be controlled and maintained by an *organization* (3.2.1) and the medium on which it is contained

Note 1 to entry: Documented information can be in any format and media and from any source.

Note 2 to entry: Documented information can refer to:

— the *management system* (3.5.3), including re-

れらの組合せがあり得る．

注記 2　文書の一式，例えば，仕様書及び記録は“文書類”と呼ばれることが多い．

注記 3　ある**要求事項**（**3.6.4**）（例えば，読むことができるという要求事項）は全ての種類の文書に関係するが，仕様書（例えば，改訂管理を行うという要求事項）及び記録（例えば，検索できるという要求事項）に対しては別の要求事項があることがある．

3.8.6

文書化した情報（documented information）

組織（**3.2.1**）が管理し，維持するよう要求されている**情報**（**3.8.2**），及びそれが含まれている媒体．

注記 1　文書化した情報は，あらゆる形式及び媒体の形をとることができ，あらゆる情報源から得ることができる．

注記 2　文書化した情報には，次に示すものがあり得る．

— 関連する**プロセス**（**3.4.1**）を含

lated *processes* (3.4.1);

— information created in order for the organization to operate (documentation);
— evidence of results achieved (*records* (3.8.10)).

Note 3 to entry: This constitutes one of the common terms and core definitions for ISO management system standards given in Annex SL of the Consolidated ISO Supplement to the ISO/IEC Directives, Part 1.

3.8.7

specification

document (3.8.5) stating *requirements* (3.6.4)

EXAMPLE *Quality manual* (3.8.8), *quality plan* (3.8.9), technical drawing, procedure document, work instruction.

Note 1 to entry: A specification can be related to activities (e.g. procedure document, *process* (3.4.1) specification and *test* (3.11.8) specification), or *products* (3.7.6) (e.g. product specification, *performance* (3.7.8) specification and drawing).

むマネジメントシステム（**3.5.3**）

— 組織の運用のために作成された情報（文書類）

— 達成された結果の証拠［**記録**（**3.8.10**）］

注記 3 この用語及び定義は，ISO/IEC 専門業務用指針—第 1 部：統合版 ISO 補足指針の**附属書 SL** に示された ISO マネジメントシステム規格の共通用語及び中核となる定義の一つを成す．

3.8.7

仕様書（specification）

要求事項（**3.6.4**）を記述した**文書**（**3.8.5**）．

例 **品質マニュアル**（**3.8.8**），**品質計画書**（**3.8.9**），技術図面，手順を記した文書，作業指示書

注記 1 仕様書には，活動に関するもの［**例** 手順を記した文書，**プロセス**（**3.4.1**）仕様書及び**試験**（**3.11.8**）仕様書］，又は**製品**（**3.7.6**）に関するもの［**例** 製品仕様書，**パフォーマンス**（**3.7.8**）

Note 2 to entry: It can be that, by stating requirements, a specification additionally is stating results achieved by *design and development* (3.4.8) and thus in some cases can be used as a *record* (3.8.10).

3.8.8

quality manual

specification (3.8.7) for the *quality management system* (3.5.4) of an *organization* (3.2.1)

Note 1 to entry: Quality manuals can vary in detail and format to suit the size and complexity of an individual *organization* (3.2.1).

3.8.9

quality plan

specification (3.8.7) of the *procedures* (3.4.5) and associated resources to be applied when and by whom to a specific *object* (3.6.1)

Note 1 to entry: These procedures generally in-

仕様書，図面］があり得る．

注記 2　要求事項を記述することによって，仕様書に，**設計・開発**（**3.4.8**）によって達成された結果が追加的に記述されることがある．この場合，仕様書が**記録**（**3.8.10**）として用いられることがある．

3.8.8

品質マニュアル（quality manual）

組織（**3.2.1**）の**品質マネジメントシステム**（**3.5.4**）についての**仕様書**（**3.8.7**）．

注記　個々の**組織**（**3.2.1**）の規模及び複雑さに応じて，品質マニュアルの詳細及び書式は変わり得る．

3.8.9

品質計画書（quality plan）

個別の**対象**（**3.6.1**）に対して，どの**手順**（**3.4.5**）及びどの関連する資源を，いつ誰によって適用するかについての**仕様書**（**3.8.7**）．

注記 1　通常，これらの手順には，**品質マネジ**

clude those referring to *quality management* (3.3.4) *processes* (3.4.1) and to *product* (3.7.6) and *service* (3.7.7) realization processes.

Note 2 to entry: A quality plan often makes reference to parts of the *quality manual* (3.8.8) or to procedure *documents* (3.8.5).
Note 3 to entry: A quality plan is generally one of the results of *quality planning* (3.3.5).

3.8.10
record
document (3.8.5) stating results achieved or providing evidence of activities performed

Note 1 to entry: Records can be used, for example, to formalize *traceability* (3.6.13) and to provide evidence of *verification* (3.8.12), *preventive action* (3.12.1) and *corrective action* (3.12.2).

Note 2 to entry: Generally records need not be under revision control.

メント（**3.3.4**）の**プロセス**（**3.4.1**）並びに**製品**（**3.7.6**）及び**サービス**（**3.7.7**）実現のプロセスに関連するものが含まれる.

注記 2　品質計画書は，**品質マニュアル**（**3.8.8**）又は手順を記した**文書**（**3.8.5**）を引用することが多い.

注記 3　品質計画書は，通常，**品質計画**（**3.3.5**）の結果の一つである.

3.8.10

記録（record）

達成した結果を記述した，又は実施した活動の証拠を提供する**文書**（**3.8.5**）.

注記 1　記録は，例えば，次のために使用されることがある.

— **トレーサビリティ**（**3.6.13**）を正式なものにする.

— **検証**（**3.8.12**），**予防処置**（**3.12.1**）及び**是正処置**（**3.12.2**）の証拠を提供する.

注記 2　通常，記録の改訂管理を行う必要はない.

3.8.11

project management plan

document (3.8.5) specifying what is necessary to meet the *objective(s)* (3.7.1) of the *project* (3.4.2)

Note 1 to entry: A project management plan should include or refer to the project's *quality plan* (3.8.9).

Note 2 to entry: The project management plan also includes or references such other plans as those relating to organizational structures, resources, schedule, budget, *risk* (3.7.9) *management* (3.3.3), environmental management, health and safety management, and security management, as appropriate.

[SOURCE: ISO 10006:2003, 3.7]

3.8.12

verification

confirmation, through the provision of *objective evidence* (3.8.3), that specified *requirements* (3.6.4) have been fulfilled

3.8.11

プロジェクトマネジメント計画書（project management plan）

プロジェクト（**3.4.2**）の**目標**（**3.7.1**）を満たすために必要な事項を明記している**文書**（**3.8.5**）.

注記 1　プロジェクトマネジメント計画書には，プロジェクトの**品質計画書**（**3.8.9**）を含めるか又は引用するとよい.

注記 2　プロジェクトマネジメント計画書には，また，組織構造，資源，日程表，予算，**リスク**（**3.7.9**）**マネジメント**（**3.3.3**），環境マネジメント，安全衛生マネジメント，セキュリティマネジメントなどの計画書を適宜含むか又は引用する.

（**JIS Q 10006**:2004 の **3.7** 参照）

3.8.12

検証（verification）

客観的証拠（**3.8.3**）を提示することによって，規定**要求事項**（**3.6.4**）が満たされていることを確認すること.

Note 1 to entry: The objective evidence needed for a verification can be the result of an *inspection* (3.11.7) or of other forms of *determination* (3.11.1) such as performing alternative calculations or reviewing *documents* (3.8.5).

Note 2 to entry: The activities carried out for verification are sometimes called a qualification *process* (3.4.1).

Note 3 to entry: The word "verified" is used to designate the corresponding status.

3.8.13

validation

confirmation, through the provision of *objective evidence* (3.8.3), that the *requirements* (3.6.4) for a specific intended use or application have been fulfilled

Note 1 to entry: The objective evidence needed for a validation is the result of a *test* (3.11.8) or other form of *determination* (3.11.1) such as performing alternative calculations or reviewing *documents* (3.8.5).

注記 1　検証のために必要な客観的証拠は，**検査**（**3.11.7**）の結果，又は別法による計算の実施若しくは**文書**（**3.8.5**）のレビューのような他の形の**確定**（**3.11.1**）の結果であることがある．

注記 2　検証のために行われる活動は，適格性**プロセス**（**3.4.1**）と呼ばれることがある．

注記 3　“検証済み”という言葉は，検証が済んでいる状態を示すために用いられる．

3.8.13

妥当性確認（validation）

客観的証拠（**3.8.3**）を提示することによって，特定の意図された用途又は適用に関する**要求事項**（**3.6.4**）が満たされていることを確認すること．

注記 1　妥当性確認のために必要な客観的証拠は，**試験**（**3.11.8**）の結果，又は別法による計算の実施若しくは**文書**（**3.8.5**）のレビューのような他の形の**確定**（**3.11.1**）の結果である．

Note 2 to entry: The word "validated" is used to designate the corresponding status.

Note 3 to entry: The use conditions for validation can be real or simulated.

3.8.14

configuration status accounting

formalized recording and reporting of *product configuration information* (3.6.8), the status of proposed changes and the status of the implementation of approved changes

[SOURCE: ISO 10007:2003, 3.7]

3.8.15

specific case

<quality plan> subject of the *quality plan* (3.8.9)

Note 1 to entry: This term is used to avoid repetition of "*process* (3.4.1), *product* (3.7.6), *project* (3.4.2) or *contract* (3.4.7)" within ISO 10005.

注記 2　“妥当性確認済み” という言葉は，妥当性確認が済んでいる状態を示すために用いられる．

注記 3　妥当性確認のための使用条件は，実環境の場合も，模擬の場合もある．

3.8.14

コンフィギュレーション状況の報告（configuration status accounting）

製品コンフィギュレーション情報（**3.6.8**），変更提案の状況，及び承認された変更の実施状況に関する，正式な記録及び報告．

（**ISO 10007**:2003 の **3.7** 参照）

3.8.15

個別ケース（specific case）

＜品質計画書＞**品質計画書**（**3.8.9**）が扱っている主題．

注記　この用語は，**ISO 10005** において，“**プロセス**（**3.4.1**），**製品**（**3.7.6**），**プロジェクト**（**3.4.2**）又は**契約**（**3.4.7**）” の反

[SOURCE: ISO 10005:2005, 3.10, modified — Note 1 to entry has been modified]

3.9 Terms related to customer

3.9.1

feedback

<customer satisfaction> opinions, comments and expressions of interest in a *product* (3.7.6), a *service* (3.7.7) or a complaints-handling *process* (3.4.1)

[SOURCE: ISO 10002:2014, 3.6, modified — The term "service" has been included in the definition]

3.9.2

customer satisfaction

customer's (3.2.4) perception of the degree to which the customer's expectations have been fulfilled

Note 1 to entry: It can be that the customer's ex-

復を避けるために使われている.

（**ISO 10005**:2005 の **3.10** の**注記 1** を変更.）

3.9　顧客に関する用語

3.9.1

フィードバック（feedback）

＜顧客満足＞**製品**（**3.7.6**），**サービス**（**3.7.7**）又は苦情対応**プロセス**（**3.4.1**）への意見，コメント，及び関心の表現.

（**JIS Q 10002**:2015 の **3.6** を変更. 用語“サービス”を定義に追加した.）

3.9.2

顧客満足（customer satisfaction）

顧客（**3.2.4**）の期待が満たされている程度に関する顧客の受け止め方.

注記 1　**製品**（**3.7.6**）又は**サービス**（**3.7.7**）

pectation is not known to the *organization* (3.2.1), or even to the customer in question, until the *product* (3.7.6) or *service* (3.7.7) is delivered. It can be necessary for achieving high customer satisfaction to fulfil an expectation of a customer even if it is neither stated nor generally implied or obligatory.

Note 2 to entry: *Complaints* (3.9.3) are a common indicator of low customer satisfaction but their absence does not necessarily imply high customer satisfaction.

Note 3 to entry: Even when customer *requirements* (3.6.4) have been agreed with the customer and fulfilled, this does not necessarily ensure high customer satisfaction.

[SOURCE: ISO 10004:2012, 3.3, modified — Notes have been modified]

3.9.3

complaint

<customer satisfaction> expression of dissatisfaction made to an *organization* (3.2.1), related to its

が引き渡されるまで，顧客の期待が，**組織**（**3.2.1**）に知られていない又は顧客本人も認識していないことがある．顧客の期待が明示されていない，暗黙のうちに了解されていない又は義務として要求されていない場合でも，これを満たすという高い顧客満足を達成することが必要なことがある．

注記 2　**苦情**（**3.9.3**）は，顧客満足が低いことの一般的な指標であるが，苦情がないことが必ずしも顧客満足が高いことを意味するわけではない．

注記 3　顧客**要求事項**（**3.6.4**）が顧客と合意され，満たされている場合でも，それが必ずしも顧客満足が高いことを保証するものではない．

（**ISO 10004**:2012 の **3.3** の**注記**を変更．）

3.9.3

苦情（complaint）

＜顧客満足＞**製品**（**3.7.6**）若しくは**サービス**（**3.7.7**）又は苦情対応**プロセス**（**3.4.1**）に関して，

product (3.7.6) or *service* (3.7.7), or the complaints-handling *process* (3.4.1) itself, where a response or resolution is explicitly or implicitly expected

[SOURCE: ISO 10002:2014, 3.2, modified — The term “service” has been included in the definition]

3.9.4

customer service

interaction of the *organization* (3.2.1) with the *customer* (3.2.4) throughout the life cycle of a *product* (3.7.6) or a *service* (3.7.7)

[SOURCE: ISO 10002:2014, 3.5, modified — The term “service” has been included in the definition]

3.9.5

customer satisfaction code of conduct

promises, made to *customers* (3.2.4) by an *organization* (3.2.1) concerning its behaviour, that are

組織（**3.2.1**）に対する不満足の表現であって，その対応又は解決を，明示的又は暗示的に期待しているもの.

（**JIS Q 10002**:2015 の **3.2** を変更. 用語“サービス”を定義に追加した.）

3.9.4

顧客サービス（customer service）

製品（**3.7.6**）又は**サービス**（**3.7.7**）のライフサイクル全般にわたる，**組織**（**3.2.1**）の**顧客**（**3.2.4**）との相互関係.

（**JIS Q 10002**:2015 の **3.5** を変更. 用語“サービス”を定義に追加した.）

3.9.5

顧客満足行動規範（customer satisfaction code of conduct）

顧客満足（**3.9.2**）を高めるための，**組織**（**3.2.1**）の行為に関する，**顧客**（**3.2.4**）への約束事項及び

aimed at enhanced *customer satisfaction* (3.9.2) and related provisions

Note 1 to entry: Related provisions can include *objectives* (3.7.1), conditions, limitations, contact *information* (3.8.2), and *complaints* (3.9.3) handling *procedures* (3.4.5).
Note 2 to entry: In ISO 10001:2007, the term “code” is used instead of “customer satisfaction code of conduct”.

[SOURCE: ISO 10001:2007, 3.1, modified — The term “code” has been removed as an admitted term, and Note 2 to entry has been modified]

3.9.6
dispute
<customer satisfaction> disagreement, arising from a *complaint* (3.9.3), submitted to a *DRP-provider* (3.2.7)

Note 1 to entry: Some *organizations* (3.2.1) allow their *customers* (3.2.4) to express their dissatisfaction to a DRP-provider in the first instance. In

関連規定.

注記 1 関連規定は, **目標**（**3.7.1**）, 条件, 制限, 連絡先の**情報**（**3.8.2**）, 及び**苦情**（**3.9.3**）対応**手順**（**3.4.5**）を含むことがある.

注記 2 **JIS Q 10001**:2010 では, “顧客満足行動規範” という用語の代わりに “規範” を用いている.

（**JIS Q 10001**:2010 の **3.1** を変更. 許容用語としての用語 “規範” は削除した. また, **注記 2** を変更した.）

3.9.6

紛争（dispute）

<顧客満足>**苦情**（**3.9.3**）から端を発し, **DRP 提供者**（**3.2.7**）に対して申し立てられる, 合意のない状態.

注記 **組織**（**3.2.1**）によっては, **顧客**（**3.2.4**）が DRP 提供者に対して最初から不満足を表明することを認めている. この場合,

this situation, the expression of dissatisfaction becomes a complaint when sent to the organization for a response, and becomes a dispute if not resolved by the organization without DRP-provider intervention. Many organizations prefer their customers to first express any dissatisfaction to the organization before utilizing dispute resolution external to the organization.

[SOURCE: ISO 10003:2007, 3.6, modified]

3.10 Terms related to characteristic

3.10.1

characteristic

distinguishing feature

Note 1 to entry: A characteristic can be inherent or assigned.

Note 2 to entry: A characteristic can be qualitative or quantitative.

Note 3 to entry: There are various classes of characteristic, such as the following:

a) physical (e.g. mechanical, electrical, chemical or biological characteristics);

その不満足の表明は，回答を求めて組織に送られた時点で苦情となり，また，DRP 提供者の介入なしに組織によって解決されなかった場合に紛争となる．多くの組織は，顧客が，組織の外部における紛争解決手続を利用する前に，不満足をまず組織に対して表明することを望む．

（**JIS Q 10003**:2010 の **3.6** を変更．）

3.10　特性に関する用語

3.10.1

特性（characteristic）

特徴付けている性質．

注記 1　特性は，本来備わっているもの又は付与されたもののいずれでもあり得る．

注記 2　特性は，定性的又は定量的のいずれでもあり得る．

注記 3　特性には，次に示すように様々な種類がある．

a)　**物質的**（**例**　機械的，電気的，化学的，生物学的）

b) sensory (e.g. related to smell, touch, taste, sight, hearing);

c) behavioural (e.g. courtesy, honesty, veracity);

d) temporal (e.g. punctuality, reliability, availability, continuity);

e) ergonomic (e.g. physiological characteristic, or related to human safety);

f) functional (e.g. maximum speed of an aircraft).

3.10.2

quality characteristic

inherent *characteristic* (3.10.1) of an *object* (3.6.1) related to a *requirement* (3.6.4)

Note 1 to entry: Inherent means existing in something, especially as a permanent characteristic.

Note 2 to entry: A characteristic assigned to an object (e.g. the price of an object) is not a quality characteristic of that object.

b)　**感覚的**（**例**　嗅覚，触覚，味覚，視覚，聴覚などに関するもの）

c)　**行動的**（**例**　礼儀正しさ，正直さ，誠実さ）

d)　**時間的**（**例**　時間厳守の度合い，信頼性，アベイラビリティ，継続性）

e)　**人間工学的**（**例**　生理学上の特性，人の安全に関するもの）

f)　**機能的**（**例**　飛行機の最高速度）

3.10.2

品質特性（quality characteristic）

要求事項（**3.6.4**）に関連する，**対象**（**3.6.1**）に本来備わっている**特性**（**3.10.1**）.

注記 1　“本来備わっている”とは，あるものに内在していること，特に，永久不変の特性として内在していることを意味する.

注記 2　対象に付与された特性（**例**　対象の価格）は，その対象の品質特性ではない.

3.10.3

human factor

characteristic (3.10.1) of a person having an impact on an *object* (3.6.1) under consideration

Note 1 to entry: Characteristics can be physical, cognitive or social.

Note 2 to entry: Human factors can have a significant impact on a *management system* (3.5.3).

3.10.4

competence

ability to apply knowledge and skills to achieve intended results

Note 1 to entry: Demonstrated competence is sometimes referred to as qualification.

Note 2 to entry: This constitutes one of the common terms and core definitions for ISO management system standards given in Annex SL of the Consolidated ISO Supplement to the ISO/IEC Directives, Part 1. The original definition has been modified by adding Note 1 to entry.

3.10.3

人的要因（human factor）

考慮の**対象**（**3.6.1**）に影響を与える，人の**特性**（**3.10.1**）.

注記 1　特性には，物理的，認知的又は社会的なものがあり得る.

注記 2　人的要因は，**マネジメントシステム**（**3.5.3**）に重要な影響を与え得る.

3.10.4

力量（competence）

意図した結果を達成するために，知識及び技能を適用する能力.

注記 1　実証された力量は，適格性ともいう.

注記 2　この用語及び定義は，ISO/IEC 専門業務用指針—第 1 部：統合版 ISO 補足指針の**附属書 SL** に示された ISO マネジメントシステム規格の共通用語及び中核となる定義の一つを成す．元の定義にない**注記 1** を追加した.

3.10.5

metrological characteristic

characteristic (3.10.1) which can influence the results of *measurement* (3.11.4)

Note 1 to entry: *Measuring equipment* (3.11.6) usually has several metrological characteristics.

Note 2 to entry: Metrological characteristics can be the subject of calibration.

3.10.6

configuration

interrelated functional and physical *characteristics* (3.10.1) of a *product* (3.7.6) or *service* (3.7.7) defined in *product configuration information* (3.6.8)

[SOURCE: ISO 10007:2003, 3.3, modified — The term "service" has been included in the definition]

3.10.5

計量特性（metrological characteristic）

測定（**3.11.4**）結果に影響を与え得る**特性**（**3.10.1**）.

注記 1　**測定機器**（**3.11.6**）は，通常複数の計量特性をもつ.

注記 2　計量特性は，校正の対象となり得る.

3.10.6

コンフィギュレーション（configuration）

製品コンフィギュレーション情報（**3.6.8**）で定義された，相互に関連する機能的及び物理的な**製品**（**3.7.6**）又は**サービス**（**3.7.7**）の**特性**（**3.10.1**）.

（**ISO 10007**:2003 の **3.3** を変更．用語“サービス”を定義に追加した.）

3.10.7

configuration baseline

approved *product configuration information* (3.6.8) that establishes the *characteristics* (3.10.1) of a *product* (3.7.6) or *service* (3.7.7) at a point in time that serves as reference for activities throughout the life cycle of the product or service

[SOURCE: ISO 10007:2003, 3.4, modified — The term "service" has been included in the definition]

3.11 Terms related to determination

3.11.1

determination

activity to find out one or more *characteristics* (3.10.1) and their characteristic values

3.11.2

review

determination (3.11.1) of the suitability, adequacy or *effectiveness* (3.7.11) of an *object* (3.6.1) to achieve established *objectives* (3.7.1)

3.10.7

コンフィギュレーションベースライン（configuration baseline）

製品（**3.7.6**）又は**サービス**（**3.7.7**）のライフサイクルを通して活動の基準となるある時点においての製品又はサービスの**特性**（**3.10.1**）を定める，承認された**製品コンフィギュレーション情報**（**3.6.8**）.

（**ISO 10007**:2003 の **3.4** を変更. 用語“サービス”を定義に追加した.）

3.11　確定に関する用語

3.11.1

確定（determination）

一つ又は複数の**特性**（**3.10.1**），及びその特性の値を見出すための活動.

3.11.2

レビュー（review）

設定された**目標**（**3.7.1**）を達成するための**対象**（**3.6.1**）の適切性，妥当性又は**有効性**（**3.7.11**）の**確定**（**3.11.1**）.

EXAMPLE Management review, *design and development* (3.4.8) review, review of *customer* (3.2.4) *requirements* (3.6.4), review of *corrective action* (3.12.2) and peer review.

Note 1 to entry: Review can also include the determination of *efficiency* (3.7.10).

3.11.3

monitoring

determining (3.11.1) the status of a *system* (3.5.1), a *process* (3.4.1), a *product* (3.7.6), a *service* (3.7.7), or an activity

Note 1 to entry: For the determination of the status there can be a need to check, supervise or critically observe.

Note 2 to entry: Monitoring is generally a determination of the status of an *object* (3.6.1), carried out at different stages or at different times.

Note 3 to entry: This constitutes one of the common terms and core definitions for ISO management system standards given in Annex SL of the Consolidated ISO Supplement to the ISO/IEC Di-

例　マネジメントレビュー，**設計・開発**（**3.4.8**）のレビュー，**顧客**（**3.2.4**）**要求事項**（**3.6.4**）のレビュー，**是正処置**（**3.12.2**）のレビュー，同等性レビュー

注記　レビューには，**効率**（**3.7.10**）の確定を含むこともある．

3.11.3

監視（monitoring）

システム（**3.5.1**），**プロセス**（**3.4.1**），**製品**（**3.7.6**），**サービス**（**3.7.7**）又は活動の状況を**確定**（**3.11.1**）すること．

注記 1　状況の確定のために，点検，監督又は注意深い観察が必要な場合もある．

注記 2　監視は，通常，異なる段階又は異なる時間において行われる，**対象**（**3.6.1**）の状況の確定である．

注記 3　この用語及び定義は，ISO/IEC 専門業務用指針―第 1 部：統合版 ISO 補足指針の**附属書 SL** に示された ISO マネジメントシステム規格の共通用語

rectives, Part 1. The original definition and Note 1 to entry have been modified, and Note 2 to entry has been added.

3.11.4

measurement

process (3.4.1) to determine a value

Note 1 to entry: According to ISO 3534-2, the value determined is generally the value of a quantity.
Note 2 to entry: This constitutes one of the common terms and core definitions for ISO management system standards given in Annex SL of the Consolidated ISO Supplement to the ISO/IEC Directives, Part 1. The original definition has been modified by adding Note 1 to entry.

3.11.5

measurement process

set of operations to determine the value of a quantity

及び中核となる定義の一つを成す．元の定義及び**注記 1** を変更し，元の定義にない**注記 2** を追加した．

3.11.4

測定（measurement）

値を確定する**プロセス**（**3.4.1**）．

注記 1　**JIS Z 8101-2** によれば，確定される値は，一般に，量の値である．

注記 2　この用語及び定義は，ISO/IEC 専門業務用指針―第 1 部：統合版 ISO 補足指針の**附属書 SL** に示された ISO マネジメントシステム規格の共通用語及び中核となる定義の一つを成す．元の定義にない**注記 1** を追加した．

3.11.5

測定プロセス（measurement process）

ある量の値を確定するための一連の操作．

3.11.6

measuring equipment

measuring instrument, software, measurement standard, reference material or auxiliary apparatus or combination thereof necessary to realize a *measurement process* (3.11.5)

3.11.7

inspection

determination (3.11.1) of *conformity* (3.6.11) to specified *requirements* (3.6.4)

Note 1 to entry: If the result of an inspection shows conformity, it can be used for purposes of *verification* (3.8.12).

Note 2 to entry: The result of an inspection can show conformity or *nonconformity* (3.6.9) or a degree of conformity.

3.11.8

test

determination (3.11.1) according to *requirements* (3.6.4) for a specific intended use or application

3.11.6

測定機器（measuring equipment）

測定プロセス（**3.11.5**）の実現に必要な計器，ソフトウェア，計量標準，標準物質若しくは補助装置，又はそれらの組合せ．

3.11.7

検査（inspection）

規定**要求事項**（**3.6.4**）への**適合**（**3.6.11**）を**確定**（**3.11.1**）すること．

注記 1　検査の結果が適合を示している場合，その結果を**検証**（**3.8.12**）のために使用することができる．

注記 2　検査の結果は，適合若しくは**不適合**（**3.6.9**），又は適合の程度を示すことがある．

3.11.8

試験（test）

特定の意図した用途又は適用に関する**要求事項**（**3.6.4**）に従って，**確定**（**3.11.1**）すること．

Note 1 to entry: If the result of a test shows *conformity* (3.6.11), it can be used for purposes of *validation* (3.8.13).

3.11.9

progress evaluation

<project management> assessment of progress made on achievement of the *project* (3.4.2) *objectives* (3.7.1)

Note 1 to entry: This assessment should be carried out at appropriate points in the project life cycle across project *processes* (3.4.1), based on criteria for project processes and *product* (3.7.6) or *service* (3.7.7).

Note 2 to entry: The results of progress evaluations can lead to revision of the *project management plan* (3.8.11).

[SOURCE: ISO 10006:2003, 3.4, modified — Notes to entry have been modified]

注記　試験の結果が**適合**（**3.6.11**）を示している場合，その結果を**妥当性確認**（**3.8.13**）のために使用することができる．

3.11.9

進捗評価（progress evaluation）

＜プロジェクトマネジメント＞**プロジェクト**（**3.4.2**）の**目標**（**3.7.1**）の達成に関してなされる進捗の査定．

注記 1　この査定は，プロジェクトの種々の**プロセス**（**3.4.1**）を通じて，プロジェクトのライフサイクルの適切な時点で，プロジェクトのプロセス及び**製品**（**3.7.6**）又は**サービス**（**3.7.7**）の基準に基づいて，実施するとよい．

注記 2　進捗評価の結果によって，**プロジェクトマネジメント計画書**（**3.8.11**）の改訂が必要になることがある．

（**JIS Q 10006**:2004 の **3.4** の**参考**を変更．）

3.12 Terms related to action

3.12.1

preventive action

action to eliminate the cause of a potential *nonconformity* (3.6.9) or other potential undesirable situation

Note 1 to entry: There can be more than one cause for a potential nonconformity.

Note 2 to entry: Preventive action is taken to prevent occurrence whereas *corrective action* (3.12.2) is taken to prevent recurrence.

3.12.2

corrective action

action to eliminate the cause of a *nonconformity* (3.6.9) and to prevent recurrence

Note 1 to entry: There can be more than one cause for a nonconformity.

Note 2 to entry: Corrective action is taken to prevent recurrence whereas *preventive action* (3.12.1) is taken to prevent occurrence.

Note 3 to entry: This constitutes one of the com-

3.12　処置に関する用語

3.12.1

予防処置（preventive action）

起こり得る**不適合**（**3.6.9**）又はその他の起こり得る望ましくない状況の原因を除去するための処置.

注記 1　起こり得る不適合には，複数の原因がある場合がある.

注記 2　**是正処置**（**3.12.2**）は再発を防止するためにとるのに対し，予防処置は発生を未然に防止するためにとる.

3.12.2

是正処置（corrective action）

不適合（**3.6.9**）の原因を除去し，再発を防止するための処置.

注記 1　不適合には，複数の原因がある場合がある.

注記 2　**予防処置**（**3.12.1**）は発生を未然に防止するためにとるのに対し，是正処置は再発を防止するためにとる.

注記 3　この用語及び定義は，ISO/IEC 専門

mon terms and core definitions for ISO management system standards given in Annex SL of the Consolidated ISO Supplement to the ISO/IEC Directives, Part 1. The original definition has been modified by adding Notes 1 and 2 to entry.

3.12.3

correction

action to eliminate a detected *nonconformity* (3.6.9)

Note 1 to entry: A correction can be made in advance of, in conjunction with or after a *corrective action* (3.12.2).

Note 2 to entry: A correction can be, for example, *rework* (3.12.8) or *regrade* (3.12.4).

3.12.4

regrade

alteration of the *grade* (3.6.3) of a *nonconforming* (3.6.9) *product* (3.7.6) or *service* (3.7.7) in order to make it conform to *requirements* (3.6.4) differing from the initial requirements

業務用指針—第1部：統合版ISO補足指針の**附属書SL**に示されたISOマネジメントシステム規格の共通用語及び中核となる定義の一つを成す．元の定義にない**注記1**及び**注記2**を追加した．

3.12.3

修正（correction）

検出された**不適合**（**3.6.9**）を除去するための処置．

注記1　**是正処置**（**3.12.2**）に先立って，是正処置と併せて，又は是正処置の後に，修正が行われることもある．

注記2　修正として，例えば，**手直し**（**3.12.8**），**再格付け**（**3.12.4**）がある．

3.12.4

再格付け（regrade）

当初の要求事項とは異なる**要求事項**（**3.6.4**）に適合するように，**不適合**（**3.6.9**）となった**製品**（**3.7.6**）又は**サービス**（**3.7.7**）の**等級**（**3.6.3**）を変更すること．

3.12.5

concession

permission to use or *release* (3.12.7) a *product* (3.7.6) or *service* (3.7.7) that does not conform to specified *requirements* (3.6.4)

Note 1 to entry: A concession is generally limited to the delivery of products and services that have *nonconforming* (3.6.9) *characteristics* (3.10.1) within specified limits and is generally given for a limited quantity of products and services or period of time, and for a specific use.

3.12.6

deviation permit

permission to depart from the originally specified *requirements* (3.6.4) of a *product* (3.7.6) or *service* (3.7.7) prior to its realization

Note 1 to entry: A deviation permit is generally given for a limited quantity of products and services or period of time, and for a specific use.

3.12.5

特別採用（concession）

規定**要求事項**（**3.6.4**）に適合していない**製品**（**3.7.6**）又は**サービス**（**3.7.7**）の使用又は**リリース**（**3.12.7**）を認めること．

注記　通常，特別採用は，特定の限度内で**不適合**（**3.6.9**）となった**特性**（**3.10.1**）をもつ製品及びサービスを引き渡す場合に限定される．また，製品及びサービスの数量又は期間を限定し，また，特定の用途に対して与えられる．

3.12.6

逸脱許可（deviation permit）

製品（**3.7.6**）又は**サービス**（**3.7.7**）の当初の規定**要求事項**（**3.6.4**）からの逸脱を，製品又はサービスの実現に先立ち認めること．

注記　逸脱許可は，一般に，製品及びサービスの数量又は期間を限定し，また，特定の用途に対して与えられる．

3.12.7

release

permission to proceed to the next stage of a *process* (3.4.1) or the next process

Note 1 to entry: In English, in the context of software and *documents* (3.8.5), the word "release" is frequently used to refer to a version of the software or the document itself.

3.12.8

rework

action on a *nonconforming* (3.6.9) *product* (3.7.6) or *service* (3.7.7) to make it conform to the *requirements* (3.6.4)

Note 1 to entry: Rework can affect or change parts of the nonconforming product or service.

3.12.9

repair

action on a *nonconforming* (3.6.9) *product* (3.7.6) or *service* (3.7.7) to make it acceptable for the in-

3.12.7

リリース（release）

プロセス（**3.4.1**）の次の段階又は次のプロセスに進めることを認めること.

注記　ソフトウェア及び**文書**（**3.8.5**）の分野では,“リリース”という言葉を, ソフトウェア自体又は文書自体の版を指すために使うことが多い.

3.12.8

手直し（rework）

要求事項（**3.6.4**）に適合させるため, **不適合**（**3.6.9**）となった**製品**（**3.7.6**）又は**サービス**（**3.7.7**）に対してとる処置.

注記　手直しは, 不適合となった製品又はサービスの部分に影響を及ぼす又は部分を変更することがある.

3.12.9

修理（repair）

意図された用途に対して受入れ可能とするため, **不適合**（**3.6.9**）となった**製品**（**3.7.6**）又は**サービ**

tended use

Note 1 to entry: A successful repair of a nonconforming product or service does not necessarily make the product or service conform to the *requirements* (3.6.4). It can be that in conjunction with a repair a *concession* (3.12.5) is required.

Note 2 to entry: Repair includes remedial action taken on a previously conforming product or service to restore it for use, for example as part of maintenance.

Note 3 to entry: Repair can affect or change parts of the nonconforming product or service.

3.12.10

scrap

action on a *nonconforming* (3.6.9) *product* (3.7.6) or *service* (3.7.7) to preclude its originally intended use

EXAMPLE Recycling, destruction.

ス（**3.7.7**）に対してとる処置.

注記 1　不適合となった製品又はサービスの修理が成功しても，必ずしも製品又はサービスが**要求事項**（**3.6.4**）に適合するとは限らない．修理と併せて，**特別採用**（**3.12.5**）が必要となることがある.

注記 2　修理には，例えば，保守の一環として，以前は適合していた製品又はサービスを使用できるように元に戻す，修復するためにとる処置を含む.

注記 3　修理は，不適合となった製品又はサービスの部分に影響を及ぼす又は部分を変更することがある.

3.12.10

スクラップ（scrap）

当初の意図していた使用を不可能にするため，**不適合**（**3.6.9**）となった**製品**（**3.7.6**）又は**サービス**（**3.7.7**）に対してとる処置.

例　再資源化，破壊

Note 1 to entry: In a nonconforming service situation, use is precluded by discontinuing the service.

3.13 Terms related to audit

3.13.1

audit

systematic, independent and documented *process* (3.4.1) for obtaining *objective evidence* (3.8.3) and evaluating it objectively to determine the extent to which the *audit criteria* (3.13.7) are fulfilled

Note 1 to entry: The fundamental elements of an audit include the *determination* (3.11.1) of the *conformity* (3.6.11) of an *object* (3.6.1) according to a *procedure* (3.4.5) carried out by personnel not being responsible for the object audited.
Note 2 to entry: An audit can be an internal audit (first party), or an external audit (second party or third party), and it can be a *combined audit* (3.13.2) or a *joint audit* (3.13.3).

Note 3 to entry: Internal audits, sometimes called

注記　サービスにおけるスクラップとは，当該サービスが不適合の場合に，そのサービスを中止することによって，その利用を不可能にすることである．

3.13　監査に関する用語

3.13.1

監査（audit）

監査基準（**3.13.7**）が満たされている程度を判定するために，**客観的証拠**（**3.8.3**）を収集し，それを客観的に評価するための，体系的で，独立し，文書化した**プロセス**（**3.4.1**）．

注記 1　監査の基本的要素には，監査される**対象**（**3.6.1**）に関して責任を負っていない要員が実行する**手順**（**3.4.5**）に従った，対象の**適合**（**3.6.11**）の**確定**（**3.11.1**）が含まれる．

注記 2　監査は，内部監査（第一者）又は外部監査（第二者・第三者）のいずれでもあり得る．また，**複合監査**（**3.13.2**）又は**合同監査**（**3.13.3**）のいずれでもあり得る．

注記 3　内部監査は，第一者監査と呼ばれるこ

first-party audits, are conducted by, or on behalf of, the *organization* (3.2.1) itself for *management* (3.3.3) *review* (3.11.2) and other internal purposes, and can form the basis for an organization's declaration of conformity. Independence can be demonstrated by the freedom from responsibility for the activity being audited.

Note 4 to entry: External audits include those generally called second and third-party audits. Second party audits are conducted by parties having an interest in the organization, such as *customers* (3.2.4), or by other persons on their behalf. Third-party audits are conducted by external, independent auditing organizations such as those providing certification/registration of conformity or governmental agencies.

Note 5 to entry: This constitutes one of the common terms and core definitions for ISO management system standards given in Annex SL of the Consolidated ISO Supplement to the ISO/IEC Directives, Part 1. The original definition and Notes to entry have been modified to remove effect of circularity between audit criteria and audit evi-

ともあり，**マネジメント**（**3.3.3**）**レビュー**（**3.11.2**）及びその他の内部目的のために，その**組織**（**3.2.1**）自体又は代理人によって行われ，その組織の適合を宣言するための基礎となり得る．独立性は，監査されている活動に関する責任を負っていないことで実証することができる．

注記 4　外部監査には，一般的に第二者監査及び第三者監査と呼ばれるものが含まれる．第二者監査は，**顧客**（**3.2.4**）など，その組織に利害をもつ者又はその代理人によって行われる．第三者監査は，適合を認証・登録する機関又は政府機関のような，外部の独立した監査組織によって行われる．

注記 5　この用語及び定義は，ISO/IEC 専門業務用指針―第 1 部：統合版 ISO 補足指針の**附属書 SL** に示された ISO マネジメントシステム規格の共通用語及び中核となる定義の一つを成す．監査基準の定義と監査証拠の定義との間の循環の影響を取り除くため，元の定

dence term entries, and Notes 3 and 4 to entry have been added.

3.13.2

combined audit

audit (3.13.1) carried out together at a single *auditee* (3.13.12) on two or more *management systems* (3.5.3)

Note 1 to entry: The parts of a management system that can be involved in a combined audit can be identified by the relevant management system standards, product standards, service standards or process standards being applied by the *organization* (3.2.1).

3.13.3

joint audit

audit (3.13.1) carried out at a single *auditee* (3.13.12) by two or more auditing *organizations* (3.2.1)

義及び注記を変更した．また，**注記 3** 及び**注記 4** を追加した．

3.13.2

複合監査（combined audit）

一つの**被監査者**（**3.13.12**）において，複数の**マネジメントシステム**（**3.5.3**）を同時に**監査**（**3.13.1**）すること．

注記　複合監査に含め得るマネジメントシステムの部分は，**組織**（**3.2.1**）が適用している関連するマネジメントシステム規格，製品規格，サービス規格又はプロセス規格によって特定することができる．

3.13.3

合同監査（joint audit）

複数の監査する**組織**（**3.2.1**）が一つの**被監査者**（**3.13.12**）を**監査**（**3.13.1**）すること．

3.13.4

audit programme

set of one or more *audits* (3.13.1) planned for a specific time frame and directed towards a specific purpose

[SOURCE: ISO 19011:2011, 3.13, modified]

3.13.5

audit scope

extent and boundaries of an *audit* (3.13.1)

Note 1 to entry: The audit scope generally includes a description of the physical locations, organizational units, activities and *processes* (3.4.1).

[SOURCE: ISO 19011:2011, 3.14, modified — Note to entry has been modified]

3.13.6

audit plan

description of the activities and arrangements for an *audit* (3.13.1)

3.13.4

監査プログラム（audit programme）

特定の目的に向けた，決められた期間内で実行するように計画された一連の**監査**（**3.13.1**）．

（**JIS Q 19011**:2012 の **3.13** を変更．）

3.13.5

監査範囲（audit scope）

監査（**3.13.1**）の及ぶ領域及び境界．

注記　監査範囲は，一般に，場所，組織単位，活動及び**プロセス**（**3.4.1**）を示すものを含む．

（**JIS Q 19011**:2012 の **3.14** の**注記**を変更．）

3.13.6

監査計画（audit plan）

監査（**3.13.1**）のための活動及び手配事項を示すもの．

[SOURCE: ISO 19011:2011, 3.15]

3.13.7

audit criteria

set of *policies* (3.5.8), *procedures* (3.4.5) or *requirements* (3.6.4) used as a reference against which *objective evidence* (3.8.3) is compared

[SOURCE: ISO 19011:2011, 3.2, modified — The term "audit evidence" has been replaced by "objective evidence"]

3.13.8

audit evidence

records, statements of fact or other information, which are relevant to the *audit criteria* (3.13.7) and verifiable

[SOURCE: ISO 19011:2011, 3.3, modified — Note to entry has been deleted]

3.13.9

audit findings

results of the evaluation of the collected *audit evi-*

（**JIS Q 19011**:2012 の **3.15** 参照）

3.13.7

監査基準（audit criteria）

客観的証拠（**3.8.3**）と比較する基準として用いる一連の**方針**（**3.5.8**），**手順**（**3.4.5**）又は**要求事項**（**3.6.4**）.

（**JIS Q 19011**:2012 の **3.2** を変更．用語“監査証拠”を“客観的証拠”に置き換えた.）

3.13.8

監査証拠（audit evidence）

監査基準（**3.13.7**）に関連し，かつ，検証できる，記録，事実の記述又はその他の情報.

（**JIS Q 19011**:2012 の **3.3** を変更．**注記**を削除した.）

3.13.9

監査所見（audit findings）

収集された**監査証拠**（**3.13.8**）を，**監査基準**

dence (3.13.8) against *audit criteria* (3.13.7)

Note 1 to entry: Audit findings indicate *conformity* (3.6.11) or *nonconformity* (3.6.9).
Note 2 to entry: Audit findings can lead to the identification of opportunities for *improvement* (3.3.1) or recording good practices.
Note 3 to entry: In English, if the *audit criteria* (3.13.7) are selected from *statutory requirements* (3.6.6) or *regulatory requirements* (3.6.7), the audit finding can be called compliance or non-compliance.

[SOURCE: ISO 19011:2011, 3.4, modified — Note 3 to entry has been modified]

3.13.10

audit conclusion

outcome of an *audit* (3.13.1), after consideration of the audit objectives and all *audit findings* (3.13.9)

[SOURCE: ISO 19011:2011, 3.5]

(3.13.7) に対して評価した結果.

注記 1　監査所見は, **適合 (3.6.11)** 又は**不適合 (3.6.9)** を示す.

注記 2　監査所見は, **改善 (3.3.1)** の機会の特定又は優れた実践事例の記録を導き得る.

注記 3　**監査基準 (3.13.7)** が**法令要求事項 (3.6.6)** 又は**規制要求事項 (3.6.7)** から選択される場合, 監査所見は "遵守" 又は "不遵守" と呼ばれることがある.

(**JIS Q 19011**:2012 の **3.4** を変更. **注記 3** を変更した.)

3.13.10

監査結論 (audit conclusion)

監査 (3.13.1) 目的及び全ての**監査所見 (3.13.9)** を考慮した上での, 監査の結論.

(**JIS Q 19011**:2012 の **3.5** 参照)

3.13.11

audit client

organization (3.2.1) or person requesting an *audit* (3.13.1)

[SOURCE: ISO 19011:2011, 3.6, modified — Note to entry has been deleted]

3.13.12

auditee

organization (3.2.1) being audited

[SOURCE: ISO 19011:2011, 3.7]

3.13.13

guide

<audit> person appointed by the *auditee* (3.13.12) to assist the *audit team* (3.13.14)

[SOURCE: ISO 19011:2011, 3.12]

3.13.14

audit team

one or more persons conducting an *audit* (3.13.1),

3.13.11

監査依頼者（audit client）

監査（**3.13.1**）を要請する**組織**（**3.2.1**）又は個人.

（**JIS Q 19011**:2012 の **3.6** を変更．**注記**を削除した.）

3.13.12

被監査者（auditee）

監査される**組織**（**3.2.1**）.

（**JIS Q 19011**:2012 の **3.7** 参照）

3.13.13

案内役（guide）

＜監査＞**監査チーム**（**3.13.14**）を手助けするために，**被監査者**（**3.13.12**）によって指名された人.

（**JIS Q 19011**:2012 の **3.12** 参照）

3.13.14

監査チーム（audit team）

監査（**3.13.1**）を行う個人又は複数の人．必要な

supported if needed by *technical experts* (3.13.16)

Note 1 to entry: One *auditor* (3.13.15) of the audit team is appointed as the audit team leader.

Note 2 to entry: The audit team can include auditors-in-training.

[SOURCE: ISO 19011:2011, 3.9, modified]

3.13.15

auditor

person who conducts an *audit* (3.13.1)

[SOURCE: ISO 19011:2011, 3.8]

3.13.16

technical expert

<audit> person who provides specific knowledge or expertise to the *audit team* (3.13.14)

Note 1 to entry: Specific knowledge or expertise relates to the *organization* (3.2.1), the *process* (3.4.1) or activity to be audited, or language or

場合は，**技術専門家**（**3.13.16**）による支援を受ける．

注記 1　監査チームの中の一人の**監査員**（**3.13.15**）は，監査チームリーダーに指名される．

注記 2　監査チームには，訓練中の監査員を含めることができる．

（**JIS Q 19011**:2012 の **3.9** を変更．）

3.13.15

監査員（auditor）

監査（**3.13.1**）を行う人．

（**JIS Q 19011**:2012 の **3.8** 参照）

3.13.16

技術専門家（technical expert）

<監査>**監査チーム**（**3.13.14**）に特定の知識又は専門的技術を提供する人．

注記 1　特定の知識又は専門的技術とは，監査される**組織**（**3.2.1**），**プロセス**（**3.4.1**）若しくは活動に関係するもの，又は言

culture.

Note 2 to entry: A technical expert does not act as an *auditor* (3.13.15) in the *audit team* (3.13.14).

[SOURCE: ISO 19011:2011, 3.10, modified — Note 1 to entry has been modified]

3.13.17

observer

<audit> person who accompanies the *audit team* (3.13.14) but does not act as an *auditor* (3.13.15)

Note 1 to entry: An observer can be a member of the *auditee* (3.13.12), a regulator or other *interested party* (3.2.3) who witnesses the *audit* (3.13.1).

[SOURCE: ISO 19011:2011, 3.11, modified — The verb "audit" has been removed from the definition; Note to entry has been modified]

語若しくは文化に関係するものである.

注記 2　技術専門家は，**監査チーム**（**3.13.14**）の**監査員**（**3.13.15**）としての行動はしない.

（**JIS Q 19011**:2012 の **3.10** を変更．**注記 1** を変更した.）

3.13.17

オブザーバ（observer）

＜監査＞**監査チーム**（**3.13.14**）に同行するが，**監査員**（**3.13.15**）として行動しない人.

注記　オブザーバは，**監査**（**3.13.1**）に立ち会う**被監査者**（**3.13.12**）の一員，規制当局又はその他の**利害関係者**（**3.2.3**）の場合がある.

[**JIS Q 19011**:2012 の **3.11** を変更．動詞“監査を行う”（audit）を定義から取り除いた．**注記**も変更した.]

索引（五十音順）

【あ行】

アウトプット（output）355
案内役（guide）433
逸脱許可（deviation permit）415
インフラストラクチャ（infrastructure）323
運営管理（management）299
オブザーバ（observer）437

【か行】

改善（improvement）297
外部委託する（outsource）317
外部供給者（external supplier）293
外部提供者（external provider）293
革新（innovation）349
確定（determination）401
活動（activity）307
監査（audit）421
——依頼者（audit client）433
——員（auditor）435
——基準（audit criteria）429
——計画（audit plan）427
——結論（audit conclusion）431

——証拠（audit evidence）429

——所見（audit findings）429

——チーム（audit team）433

——範囲（audit scope）427

——プログラム（audit programme）427

監視（monitoring）403

技術専門家（technical expert）435

規制要求事項（regulatory requirement）341

客観的証拠（objective evidence）367

協会（association）297

供給者（supplier）291

記録（record）377

苦情（complaint）387

継続的改善（continual improvement）299

計測マネジメントシステム（measurement management system）329

契約（contract）319

計量確認（metrological confirmation）327

計量機能（metrological function）297

計量特性（metrological characteristic）399

欠陥（defect）343

検査（inspection）407

検証（verification）379

合同監査（joint audit）425

項目（item）333

効率（efficiency）365

440
顧客（customer）291
——サービス（customer service）389
——満足（customer satisfaction）385
——満足行動規範（customer satisfaction code of conduct）389
個別ケース（specific case）383
コンフィギュレーション（configuration）399
——管理（configuration management）305
——機関（configuration authority）283
——決定委員会（dispositioning authority）283
——状況の報告（configuration status accounting）383
——対象（configuration object）309
——統制委員会（configuration control board）283
——ベースライン（configuration baseline）401

【さ行】

サービス（service）359
再格付け（regrade）413
作業環境（work environment）325
参画（involvement）283
試験（test）407
システム（system）321
持続的成功（sustained success）353
実現能力（capability）345
実体（entity）333
使命（mission）333

修正（correction）413
修理（repair）417
仕様書（specification）373
情報（information）367
——システム（information system）369
進捗評価（progress evaluation）409
人的要因（human factor）397
スクラップ（scrap）419
ステークホルダー（stakeholder）289
成功（success）353
製品（product）357
——コンフィギュレーション情報（product configuration information）341
是正処置（corrective action）411
積極的参加（engagement）283
設計・開発（design and development）319
戦略（strategy）333
測定（measurement）405
——機器（measuring equipment）407
——プロセス（measurement process）405
組織（organization）285
——の状況（context of the organization）287

【た行】

対象（object）333
妥当性確認（validation）381

DRP 提供者（DRP-provider）293
提供者（provider）291
ディペンダビリティ（dependability）347
データ（data）367
適合（conformity）345
手順（procedure）317
手直し（rework）417
等級（grade）335
特性（characteristic）393
特別採用（concession）415
トップマネジメント（top management）279
トレーサビリティ（traceability）347

【は行】

パフォーマンス（performance）361
被監査者（auditee）433
ビジョン（vision）331
品質（quality）335
——改善（quality improvement）305
——管理（quality control）305
——計画（quality planning）303
——計画書（quality plan）375
——特性（quality characteristic）395
——方針（quality policy）331
——保証（quality assurance）303
——マニュアル（quality manual）375

——マネジメント（quality management）303
——マネジメントシステム（quality management system）325
——マネジメントシステムコンサルタント（quality management system consultant）281
——マネジメントシステムの実現（quality management system realization）315
——目標（quality objective）351
——要求事項（quality requirement）339
フィードバック（feedback）385
複合監査（combined audit）425
不適合（nonconformity）341
プロジェクト（project）313
——マネジメント（project management）309
——マネジメント計画書（project management plan）379
プロセス（process）309
文書（document）369
——化した情報（documented information）371
紛争（dispute）391
——解決者（dispute resolver）285
——解決手続提供者（dispute resolution process provider）293
変更管理（change control）307
方針（policy）329
法令要求事項（statutory requirement）341

【ま行】

マネジメント（management）299
——システム（management system）323
目標（objective）349

【や行】

有効性（effectiveness）365
要求事項（requirement）337
予防処置（preventive action）411

【ら行】

利害関係者（interested party）289
力量（competence）397
——の習得（competence acquisition）317
リスク（risk）363
リリース（release）417
レビュー（review）401

索引（アルファベット順）

【A】

activity（活動）306
association（協会）296
audit（監査）420
　—— client（監査依頼者）432
　—— conclusion（監査結論）430
　—— criteria（監査基準）428
　—— evidence（監査証拠）428
　—— findings（監査所見）428
　—— plan（監査計画）426
　—— programme（監査プログラム）426
　—— scope（監査範囲）426
　—— team（監査チーム）432
auditee（被監査者）432
auditor（監査員）434

【C】

capability（実現能力）344
change control（変更管理）306
characteristic（特性）392
combined audit（複合監査）424
competence（力量）396

446
competence acquisition（力量の習得）316
complaint（苦情）386
concession（特別採用）414
configuration（コンフィギュレーション）398
—— authority（コンフィギュレーション機関）282
—— baseline（コンフィギュレーションベースライン）400
—— control board（コンフィギュレーション統制委員会）282
—— management（コンフィギュレーション管理）304
—— object（コンフィギュレーション対象）308
—— status accounting（コンフィギュレーション状況の報告）382
conformity（適合）344
context of the organization（組織の状況）286
continual improvement（継続的改善）298
contract（契約）318
correction（修正）412
corrective action（是正処置）410
customer（顧客）290
—— satisfaction（顧客満足）384
—— satisfaction code of conduct（顧客満足行動規範）388
—— service（顧客サービス）388

【D】

data（データ）366
defect（欠陥）342
dependability（ディペンダビリティ）346
design and development（設計・開発）318
determination（確定）400
deviation permit（逸脱許可）414
dispositioning authority（コンフィギュレーション決定委員会）282
dispute（紛争）390
—— resolution process provider（紛争解決手続提供者）292
—— resolver（紛争解決者）284
document（文書）368
documented information（文書化した情報）370
DRP-provider（DRP 提供者）292

【E】

effectiveness（有効性）364
efficiency（効率）364
engagement（積極的参加）282
entity（実体）332
external provider（外部提供者）292
external supplier（外部供給者）292

【F】

feedback（フィードバック）384

【G】

grade（等級）334

guide（案内役）432

【H】

human factor（人的要因）396

【I】

improvement（改善）296

information（情報）366

　── system（情報システム）368

infrastructure（インフラストラクチャ）322

innovation（革新）348

inspection（検査）406

interested party（利害関係者）288

involvement（参画）282

item（項目）332

【J】

joint audit（合同監査）424

【M】

management（マネジメント，運営管理）298

　── system（マネジメントシステム）322

measurement（測定）404

—— management system（計測マネジメントシステム）328
—— process（測定プロセス）404
measuring equipment（測定機器）406
metrological characteristic（計量特性）398
metrological confirmation（計量確認）326
metrological function（計量機能）296
mission（使命）332
monitoring（監視）402

【N】

nonconformity（不適合）340

【O】

object（対象）332
objective（目標）348
—— evidence（客観的証拠）366
observer（オブザーバ）436
organization（組織）284
output（アウトプット）354
outsource（外部委託する）316

【P】

performance（パフォーマンス）360
policy（方針）328
preventive action（予防処置）410

450
procedure（手順）316
process（プロセス）308
product（製品）356
—— configuration information（製品コンフィギュレーション情報）340
progress evaluation（進捗評価）408
project（プロジェクト）312
—— management（プロジェクトマネジメント）308
—— management plan（プロジェクトマネジメント計画書）378
provider（提供者）290

【Q】

quality（品質）334
—— assurance（品質保証）302
—— characteristic（品質特性）394
—— control（品質管理）304
—— improvement（品質改善）304
—— management（品質マネジメント）302
—— management system（品質マネジメントシステム）324
—— management system consultant（品質マネジメントシステムコンサルタント）280
—— management system realization（品質マネジメントシステムの実現）314
—— manual（品質マニュアル）374

—— objective（品質目標）350
—— plan（品質計画書）374
—— planning（品質計画）302
—— policy（品質方針）330
—— requirement（品質要求事項）338

【R】

record（記録）376
regrade（再格付け）412
regulatory requirement（規制要求事項）340
release（リリース）416
repair（修理）416
requirement（要求事項）336
review（レビュー）400
rework（手直し）416
risk（リスク）362

【S】

scrap（スクラップ）418
service（サービス）358
specific case（個別ケース）382
specification（仕様書）372
stakeholder（ステークホルダー）288
statutory requirement（法令要求事項）340
strategy（戦略）332
success（成功）352

supplier（供給者）290
sustained success（持続的成功）352
system（システム）320

【T】

technical expert（技術専門家）434
test（試験）406
top management（トップマネジメント）278
traceability（トレーサビリティ）346

【V】

validation（妥当性確認）380
verification（検証）378
vision（ビジョン）330

【W】

work environment（作業環境）324

対訳 ISO 9001:2015（JIS Q 9001:2015）
品質マネジメントの国際規格［ポケット版］

2016 年 2 月26日　　第 1 版第 1 刷発行
2024 年 9 月12日　　　　　第10刷発行

監　　修　品質マネジメントシステム規格国内委員会
編　　者　一般財団法人 日本規格協会
発 行 者　朝日　弘
発 行 所　一般財団法人 日本規格協会
〒108-0073　東京都港区三田 3 丁目 11-28　三田 Avanti
https://www.jsa.or.jp/
振替　00160-2-195146

製　　作　日本規格協会ソリューションズ株式会社
印 刷 所　三美印刷株式会社
製作協力　株式会社群企画

ISBN978-4-542-30660-8

図　書　の　ご　案　内

対訳 ISO 19011:2018 (JIS Q 19011:2019)
マネジメントシステム監査のための指針
［ポケット版］

日本規格協会　編
新書判・304ページ　定価 7,480円（本体 6,800円＋税 10%）

ISO 19011:2018 (JIS Q 19011:2019)
マネジメントシステム監査
解説と活用方法

福丸典芳　著
A5判・264ページ　定価 4,290円（本体 3,900円＋税 10%）

ISO 9001:2015
内部監査の実際［第4版］

上月宏司　著
A5判・180ページ　定価 2,420円（本体 2,200円＋税 10%）

2015年版対応　中小企業のための
ISO 9001 内部監査指摘ノウハウ集

ISO 9001 内部監査指摘ノウハウ集編集委員会　編
編集委員長　福丸典芳
A5判・150ページ　定価 2,640円（本体 2,400円＋税 10%）

2015年版対応　ISO 9001/14001
内部監査のチェックポイント 222
有効で本質的なマネジメントシステムへの改善

国府保周　著
A5判・348ページ　定価 4,400円（本体 4,000円＋税 10%）

ISO 9001/14001
規格要求事項と審査の落とし穴からの脱出
思い込みと誤解はどこから生まれたか

国府保周　著
A5判・246ページ　定価 2,750円（本体 2,500円＋税 10%）

日本規格協会　https://webdesk.jsa.or.jp/